THREE ROADS TO
QUANTUM GRAVITY

science masters

Other books by Lee Smolin include

The Life of the Cosmos

THREE ROADS TO
QUANTUM GRAVITY

LEE SMOLIN

BASIC

BOOKS

A Member of the Perseus Books Group

To my parents
Pauline and Michael

Published by Basic Books,
A Member of the Perseus Books Group

First Published in Great Britian in 2000 by
Weidenfeld and Nicolson

Typeset at the Spartan Press Ltd
Lymington, Hants

A CIP record for this book is available from the Library of Congress.
ISBN 0-465-07836-2

02 03 04 / 10 9 8 7 6 5 4 3

CONTENTS

ACKNOWLEDGMENTS

I must thank first of all the friends and collaborators who formed the community within which I learned most of what I know about quantum gravity. If I had not had the luck to know Julian Barbour, Louis Crane, John Dell, Ted Jacobson and Carlo Rovelli, I doubt that I would have got very far with this subject at all. They will certainly see their ideas and views represented here. Fotini Markopoulou-Kalamara is responsible for changing my views on several important aspects of quantum gravity over the last several years, which got me out of space and back into spacetime. Large parts of Chapters 2, 3 and 4 are suffused by her thinking, as will be clear to anyone who has read her papers. I am also indebted to Chris Isham for the several ideas of his that have seeped into my work, in some cases without my fully realizing it, and for his friendship and for the example of his life and thought. Stuart Kauffman has taught me most of what I know about how to think about self-organized systems, and I thank him for both the gift of his friendship and his willingness to walk with me in the space between our two subjects.

I am also very indebted for discussions, collaborations and encouragement to Giovanni Amelino-Camelia, Abhay Ashtekar, Per Bak, Herbert Bernstein, Roumen Borissov, Alain Connes, Michael Douglas, Laurant Friedel, Rodolfo Gambini, Murray Gell-Mann, Sameer Gupta, Eli Hawkins, Gary Horowitz, Viqar Husain, Jaron Lanier, Richard Livine, Yi Ling, Renate Loll, Seth Major, Juan Maldecena, Maya Paczuski, Roger Penrose, Jorge Pullin, Martin Rees, Mike Reisenberger,

Jurgen Renn, Kelle Stelle, Andrew Strominger, Thomas Thiemann and Edward Witten. I add a special word of appreciation for the founders of the field of quantum gravity, including Peter Bergmann, Bryce DeWitt, David Finkelstein, Charles Misner, Roger Penrose and John Archibald Wheeler. If they see many of their ideas here, it is because these are the ideas that continue to shape how we see our problem. Our work has been supported generously by the National Science Foundation, for which I especially have to thank Richard Issacson. In the last several years generous and unexpected gifts from the Jesse Phillips Foundation have made it possible to concentrate on doing science at a time when nothing was more important than the time and freedom to think and work. I am grateful also to Pennsylvania State University, and especially to my chair, Jayanth Banavar, for the supportive and stimulating home it has given me over the last six years, as well as for showing some understanding of the conflicting demands placed upon me when I found myself with three full-time jobs, as a scientist, teacher and author. The theory group at Imperial College provided a most stimulating and friendly home from home during the year's sabbatical during which this book was written.

This book would not exist were it not for the kind encouragement of John Brockman and Katinka Matson, and I am also very grateful to Peter Tallack of Weidenfeld & Nicolson, both for his encouragement and ideas and for being such a good editor, in the old-fashioned sense. Much of the clarity of the text is also due to the artistry and intelligence of John Woodruff's copy editing. Brian Eno and Michael Smolin read a draft and made invaluable suggestions for the organization of the book, which have greatly improved it. The support of friends has continued to be essential to keeping my own spirit alive, especially Saint Clair Cemin, Jaron Lanier and Donna Moylan. Finally, as always, my greatest debt is to my parents and family, not only for the gift of life but for imparting in me the desire to look beyond what is taught in school, to try to see the world as it may actually be.

Lee Smolin
London, July 2000

....

THE QUEST FOR QUANTUM GRAVITY

This book is about the simplest of all questions to ask: 'What are time and space?' This is also one of the hardest questions to answer, yet the progress of science can be measured by revolutions that produce new answers to it. We are now in the midst of such a revolution, and not one but several new ideas about space and time are being considered. This book is meant to be a report from the front. My aim is to communicate these new ideas in a language that will enable any interested reader to follow these very exciting developments.

Space and time are hard to think about because they are the backdrop to all human experience. Everything that exists exists somewhere, and nothing happens that does not happen at some time. So, just as one can live without questioning the assumptions in one's native culture, it is possible to live without asking about the nature of space and time. But there is at least a moment in every child's life when they wonder about time. Does it go on for ever? Was there a first moment? Will there be a last moment? If there was a first moment, then how was the universe created? And what happened just a moment before that? If there was no first moment, does that mean that everything has happened before? And the same for space: does it go on and on for ever? If there is an end to space, what is just on the other side of it? If there isn't an end, can one count the things in the universe?

I'm sure people have been asking these questions for as long as there have been people to ask them. I would be surprised if the people who painted the walls of their caves tens of thousands of years ago did not ask them of one another as they sat around their fires after their evening meals.

For the past hundred years or so we have known that matter is made up of atoms, and that these in turn are composed of electrons, protons and neutrons. This teaches us an important lesson – that human perception, amazing as it sometimes is, is too coarse to allow us to see the building blocks of nature directly. We need new tools to see the smallest things. Microscopes let us see the cells that we and other living things are made of, but to see atoms we must look on scales at least a thousand times smaller. We can now do this with electron microscopes. Using other tools, such as particle accelerators, we can see the nucleus of an atom, and we have even seen the quarks that make up the protons and neutrons.

All this is wonderful, but it raises still more questions. Are the electrons and the quarks the smallest possible things? Or are they themselves made up of still smaller entities? As we continue to probe, will we always find smaller things, or is there a smallest possible entity? We may wonder in the same way not only about matter but also about space: space seems continuous, but is it really? Can a volume of space be divided into as many parts as we like, or is there a smallest unit of space? Is there a smallest distance? Similarly, we want to know whether time is infinitely divisible or whether there might be a smallest possible unit of time. Is there a simplest thing that can happen?

Until about a hundred years ago there was an accepted set of answers to these questions. They made up the foundations of Newton's theory of physics. At the beginning of the twentieth century people understood that this edifice, useful as it had been for so many developments in science and engineering, was completely wrong when it came to giving answers to these fundamental questions about space and time. With the overthrow of Newtonian physics came new answers to these questions. They came from new theories: principally from Albert Einstein's theory of relativity, and from the quantum theory, invented by Neils Bohr, Werner

Heisenberg, Erwin Schrödinger, and many others. But this was only the starting point of the revolution, because neither of these two theories is complete enough to serve as a new foundation for physics. While very useful, and able to explain many things, each is incomplete and limited.

Quantum theory was invented to explain why atoms are stable, and do not instantly fall apart, as was the case for all attempts to describe the structure of atoms using Newton's physics. Quantum theory also accounts for many of the observed properties of matter and radiation. Its effects differ from those predicted by Newton's theory primarily, although not exclusively, on the scale of molecules and smaller. In contrast, general relativity is a theory of space, time and cosmology. Its predictions differ strongly from Newton's mainly on very large scales, so many of the observations that confirm general relativity come from astronomy. However, general relativity seems to break down when it is confronted by the behaviour of atoms and molecules. Equally, quantum theory seems incompatible with the description of space and time that underlies Einstein's general relativity theory. Thus, one cannot simply bring the two together to construct a single theory that would hold from the atoms up to the solar system and beyond to the whole universe.

It is not difficult to explain why it is hard to bring relativity and quantum theory together. A physical theory must be more than just a catalogue of what particles and forces exist in the world. Before we even begin to describe what we see when we look around us, we must make some assumptions about what it is that we are doing when we do science. We all dream, yet most of us have no problem distinguishing our dreams from our experiences when awake. We all tell stories, but most of us believe there is a difference between fact and fiction. As a consequence, we talk about dreams, fiction and our ordinary experience in different ways which are based on different assumptions about the relation of each to reality. These assumptions can differ slightly from person to person and from culture to culture, and they are also subject to revision by artists of all kinds. If they are not spelled out the result can be confusion and disorientation, either accidental or intended.

Similarly, physical theories differ in the basic assumptions they make about observation and reality. If we are not careful to spell them out, confusion can and will occur when we try to compare descriptions of the world that come out of different theories.

In this book we shall be concerned with two very basic ways in which theories may differ. The first is in the answer they give to the question of what space and time are. Newton's theory was based on one answer to this question, general relativity on quite another. We shall see shortly what these were, but the important fact is that Einstein altered forever our understanding of space and time.

Another way in which theories may differ is in how observers are believed to be related to the system they observe. There must be some relationship, otherwise the observers would not even be aware of the existence of the system. But different theories can and do differ strongly in the assumptions they make about the relationship between observer and observed. In particular, quantum theory makes radically different assumptions from those made by Newton about this question.

The problem is that while quantum theory changed radically the assumptions about the relationship between the observer and the observed, it accepted without alteration Newton's old answer to the question of what space and time are. Just the opposite happened with Einstein's general relativity theory, in which the concept of space and time was radically changed, while Newton's view of the relationship between observer and observed was retained. Each theory seems to be at least partly true, yet each retains assumptions from the old physics that the other contradicts.

Relativity and quantum theory were therefore just the first steps in a revolution that now, a century later, remains unfinished. To complete the revolution, we must find a single theory that brings together the insights gained from relativity and quantum theory. This new theory must somehow merge the new conception of space and time Einstein introduced with the new conception of the relationship between the observer and the observed which the quantum theory teaches us. If that does not prove possible, it must reject both and find new answers to the questions of what space and time are and what the relationship between observer and observed is.

The new theory is not yet complete, but it already has a name: it is called the quantum theory of gravity. This is because a key part of it involves extending the quantum theory, which is the basis of our understanding of atoms and the elementary particles, to a theory of gravity. Gravity is presently understood in the context of general relativity, which teaches us that gravity is actually a manifestation of the structure of space and time. This was Einstein's most surprising and most beautiful insight, and we shall have a great deal to say about it as we go along. The problem we now face is (in the jargon of fundamental physics) to *unify* Einstein's theory of general relativity with the quantum theory. The product of this unification will be a quantum theory of gravity.

When we have it, the quantum theory of gravity will provide new answers to the questions of what space and time are. But that is not all. The quantum theory of gravity will also have to be a theory of matter. It will have to contain all the insights gained over the last century into the elementary particles and the forces that govern them. It must also be a theory of cosmology. It will, when we have it, answer what now seem very mysterious questions about the origin of the universe, such as whether the big bang was the first moment of time or only a transition from a different world that existed previously. It may even help us to answer the question of whether the universe was fated to contain life, or whether our own existence is merely the consequence of a lucky accident. ✗

As we enter the twenty-first century, there is no more challenging problem in science than the completion of this theory. You may wonder, as many have, whether it is too hard – whether it will remain always unsolved, in the class of impossible problems like certain mathematical problems or the nature of consciousness. It would not be surprising if, once you see the scope of the problem, you were to take this view. Many good physicists have. Twenty-five years ago, when I began to work on the quantum theory of gravity in

✗ This is the question that sepeates faith based from reason based. Besically life evolves to fill nitches— nitches do not evolve to fill life.

college, several of my teachers told me that only fools worked on this problem. At that time very few people worked seriously on quantum gravity. I don't know if they ever all got together for a dinner party, but they might have.

My advisor in graduate school, Sidney Coleman, tried to talk me into doing something else. When I persisted he told me he would give me a year to get started and that if, as he expected, I made no progress, he would assign me a more doable project in elementary particle physics. Then he did me a great favour: he asked one of the pioneers of the subject, Stanley Deser, to look after me and share my supervision. Deser had recently been one of the inventors of a new theory of gravity called supergravity, which for a few years seemed to solve many of the problems that had resisted all earlier attempts to solve them. I was also lucky during my first year at graduate school to hear lectures by someone else who had made an important contribution to the search for quantum gravity: Gerard 't Hooft. If I have not always followed either of their directions, I learned a crucial lesson from the example of their work – that it is possible to make progress on a seemingly impossible problem if one just ignores the sceptics and gets on with it. After all, atoms do fall, so the relationship between gravity and the quantum is not a problem for nature. If it is a problem for us it must be because somewhere in our thinking there is at least one, and possibly several, wrong assumptions. At the very least, these assumptions involve our concept of space and time and the connection between the observer and the observed.

It was obvious to me then that before we could find the quantum theory of gravity we first had to isolate these wrong assumptions. This made it possible to push ahead for there is an obvious strategy for rooting out false assumptions: try to construct the theory, and see where it fails. Since all the avenues that had been followed up to that time had, sooner or later, led to a dead end, there was ample work to do. It may not have inspired many people, but it was necessary work and, for a time, it was enough.

The situation now is very different. We are still not quite there, but few who work in the field doubt that we have come a long way towards our goal. The reason is that, beginning in the mid-1980s, we began to find ways of combining quantum theory and relativity that did not fail, as all previous attempts had. As a result, it is possible to say that in the last few years large parts of the puzzle have been solved.

One consequence of our having made progress is that all of a sudden our pursuit has become fashionable. The small number of pioneers who were working on the subject a few decades ago have now grown into a large community of hundreds of people who work full time on some aspect of the problem of quantum gravity. There are, indeed, so many of us that, like the jealous primates we are, we have splintered into different communities pursuing different approaches. These go under different names, such as strings, loops, twistors, non-commutative geometry and topi. This over-specialization has had unfortunate effects. In each community there are people who are sure that their approach is the only key to the problem. Sadly, most of them do not understand in any detail the main results that excite the people working on the other approaches. There are even cases in which someone taking one approach does not seem to realize that a problem they find hard has been completely solved by someone taking another approach. One consequence of this is that many people who work on some aspect of quantum gravity do not have a view of the field that is wide enough to take in all the progress that has recently been made towards its solution.

This is perhaps not so surprising – it seems not very different from the present state of cancer research or evolutionary theory. Because the problem is hard, it might be expected that, like climbers confronting a virgin peak, different people would attempt different approaches. Of course, some of these approaches will turn out to be total failures. But, at least in the case of quantum gravity, several approaches seem recently to have led to genuine discoveries about the nature of space and time.

The most compelling developments, taking place as I write, have to do with bringing together the different lessons that have been learned by following the different approaches, so that they can be incorporated into a single theory – the quantum theory of gravity. Although we do not yet have this single theory in its final form, we do know a lot about it, and this is the basis of what I shall be describing in the chapters to come.

I should warn the reader that I am by temperament a very optimistic person. My own view is that we are only a few years away from having the complete quantum theory of gravity, but I do have friends and colleagues who are more cautious. So I want to emphasize that what follows is a personal view, one that not every scientist or mathematician working on the problem of quantum gravity will endorse. I should also add that there are a few mysteries that have yet to be solved. The final stone that finishes the arch has yet to be found.

Furthermore, I must emphasize that so far it has not been possible to test any of our new theories of quantum gravity experimentally. Until very recently it was even believed that the quantum theory of gravity could not be tested with existing technology, and that it would therefore be many years into the future before the theory could be confronted with data from experimental science. However, it now appears that this pessimism may have been short-sighted. Philosophers of science such as Paul Feyerabend have stressed that new theories often suggest new kinds of experiment which may be used to test them. This is very definitely happening in quantum gravity. Very recently, new experiments have been proposed which it appears will make it possible to test at least some of the theory's predictions in the very near future. These new experiments will employ existing technology, but used in surprising ways, to study phenomena that would not have been thought, on the basis of the old theories, to have anything to do with quantum gravity. This is indeed a sign of real progress. However, we must never forget that until the experiments are performed it will always be possible that, as beautiful and compelling as the new theories may seem, they are simply wrong.

During the past few years there has been a growing sense of excitement and confidence among many of the people working on quantum gravity. It is hard to avoid the feeling that we are indeed closing in on the beast. We may not have it in our net, but it feels as if we have it cornered and we have seen, with our flashlights, a few glimpses of it.

Among the many different paths to quantum gravity, most recent traffic, and most progress, has been along three broad roads. Given that quantum gravity is supposed to arise from a unification of two theories – relativity and quantum theory – two of these paths are perhaps not unexpected. There is the route from quantum theory, in which most of the ideas and methods used were developed first in other parts of quantum theory. Then there is the road from relativity, along which one starts with the essential principles of Einstein's theory of general relativity and seeks to modify them to include quantum phenomena. These two roads have each led to a well worked-out and partly successful theory of quantum gravity. The first road gave birth to string theory, while the second led to a seemingly different theory (although with a similar name) called loop quantum gravity.

Both loop quantum gravity and string theory agree on some of the basics. They agree that there is a physical scale on which the nature of space and time is very different from that which we observe. This scale is extremely small, far out of the reach of experiments done with even the largest particle accelerators. It may in fact be very much smaller than we have so far probed. It is usually thought to be as much as 20 orders of magnitude (i.e. a factor of 10^{20}) smaller than an atomic nucleus. However, we are not really sure at which point it is reached, and recently there have been some very imaginative suggestions that, if they bear fruit, will bring quantum gravity effects within the range of present-day experimental capabilities.

The scale where quantum gravity is necessary to describe space and time is called the Planck scale. Both string theory and loop quantum gravity are theories about what space and time are like on this tiny scale. One of the stories I shall be telling is how the pictures that each theory gives us are converging. Not everyone yet agrees, but there is more and more evidence that these different approaches are different windows into the same very tiny world.

Having said this, I should confess my own situation and bias. I was one of the first people to work on loop quantum gravity. The most exhilarating days of my life (apart from the purely personal) were those when, all of a sudden, after months of hard work, we suddenly understood one of our theory's basic lessons. The friends I did that with are friends for life, and I feel equal affection and hope for the discoveries we made. But before then I worked on string theory and, for the past four years, most of my work has been in the very fertile domain that lies between the two theories. I believe that the essential results of both string theory and loop quantum gravity are true, and the picture of the world I shall be presenting here is one that comes from taking both seriously.

Apart from string theory and loop quantum gravity, there has always been a third road. This has been taken by people who discarded both relativity and quantum theory as being too flawed and incomplete to be proper starting points. Instead, these people wrestle with the fundamental principles and attempt to fashion the new theory directly from them. While they make reference to the older theories, these people are not afraid to invent whole new conceptual worlds and mathematical formalisms. Thus, unlike the other two paths, which are trodden by communities of people each large enough to exhibit the full spectrum of human group behaviour, this third path is followed by just a few individuals, each pursuing his or her own vision, each either a prophet or a fool, who prefers that essential uncertainty to the comfort of travelling with a crowd of like-minded seekers.

The journey along the third path is driven by deep, philosophical questions such as, 'What is time?' or, 'How do we describe a universe in which we are participants?' These are not easy questions, but some of the greatest minds of our time have chosen to attack them head-on, and I believe that there has been great progress along this path too. New and, in some cases, quite surprising ideas have been discovered, which I believe are up to the task of answering these questions. I believe that they provide the conceptual framework that is allowing us to take the next step – to proceed to a quantum theory of gravity.

It has also happened that someone on this third road discovered a mathematical structure which at first seemed unconnected to anything else. Such results are often dismissed by the more conservative members of the field as having no possible connection to reality, but these critics have sometimes had to eat their words when the same structure surprisingly turns up on one of the first two roads as the answer to what seemed an otherwise intractable problem. This of course only proves that fundamental questions are hardly ever solved by accident. The people who discovered these structures are among the true heroes of this story. They include Alain Connes, David Finkelstein, Christopher Isham, Roger Penrose and Raphael Sorkin.

In this book we shall walk down all three roads. We shall discover that they are closer than they seem – linked by paths, little used and perhaps a bit overgrown, but passable nevertheless. I shall argue that, if we put together the key ideas and discoveries from all the roads, a definite picture emerges of what the world is like on the Planck scale. My intention here is to display this picture and, by doing so, to show how close we are to the solution of the problem of quantum gravity.

I have tried to aim this book at the intelligent layperson, interested in knowing what is going on at the frontiers of physics. I have not assumed any previous knowledge of relativity or quantum theory. I believe that the reader who has not read anything previously on these subjects will be able to follow this book. At the same time I have introduced ideas from relativity and quantum theory only when they are needed to explain something. I could have said much more about most of the subjects I mention, even at an introductory level. But to have included a complete introduction to these subjects would have resulted in a very long book, and this would have defeated my main goal. Fortunately, there are many good introductions to these subjects for the layperson. At the end of this book there are some suggestions for further reading for those who want to know more.

I must also emphasize that in most cases I have not given proper credit to the inventors of the ideas and discoveries I present. The knowledge we have about quantum gravity has not come out of the head of two or three neo-Einsteins. Rather, it is the result of several decades of intense effort by a large and growing community of scientists. In most cases to name only a few people would be a disservice to both the community of scientists and to the reader, as it would reinforce the myth that science is done by a few great individuals in isolation. To come anywhere near the truth, even about a small field like quantum gravity, one has to describe the contributions of scores of people. There are many more people to name than could be kept track of by the reader encountering these ideas for the first time.

For a few episodes with which I was involved enough to be confident of knowing what happened, I have told the stories of how the discoveries were made. Because people are most interesting when one tells the truth about them, in these cases I am happy to introduce some very human stories to illustrate how science actually gets done. Otherwise I have stayed away from telling the stories of who did what, for I would inevitably have got some of it wrong, in spite of having been a close observer of the subject for the last two decades.

In taking the liberty of telling a few stories I also take a risk, which is that the reader will get the impression that I believe my own work to be more important than the work of other people in the field. This is not true. Of course, I do believe in the approach I pursue in my own research, otherwise I would not have a point of view worth forming a book around. But I believe that I am also in a position to make a fair appraisal of the strengths and weaknesses of all the different approaches, not only those to which I've contributed. Above all else, I feel very privileged to be part of the community of people working on quantum gravity. If I were a real writer, skilled in the art of conveying character, I would like nothing better than to describe some of the people in this world I most admire, from whom I continue to learn, every chance I get. But given my limited skills I shall stick to a few stories about people and incidents I know very well.

When our task is done, someone will write a good history of the search for quantum gravity. Whether this will be in a few years, as I believe, or in many decades, as some of my more pessimistic colleagues expect, it will be a story in which the best human virtues, of courage, wisdom and vision, are mixed with the most ordinary sort of primate behaviour, expressed through the rituals of academic politics. I hope that story will be written in a style that celebrates both sides of our very human occupation.

Each of the following chapters is devoted to one step in our search for the quantum theory of gravity. We begin with four basic principles that determine how we approach our enquiry into the nature of space, time and the cosmos. These make up the first part, called 'Points of departure'. With this preparation we turn to the second part, 'What we have learned', in which I shall describe the main conclusions that have so far been arrived at on the three roads to quantum gravity. These combine to give us a picture of what the world is like on the smallest possible scales of space and time. From there we turn to the last part, a tour of 'The present frontiers' of the subject. We shall introduce a new principle, called the holographic principle, which may very well be the fundamental principle of quantum gravity. The next chapter is a discussion of how the different approaches to quantum gravity may be coming together into one theory which seems to have the possibility of answering, at least for the foreseeable future, our questions about the nature of space and time. I end with a reflection on the question of how the universe chose the laws of nature.

We begin at the beginning, with the first principle.

I
POINTS OF
DEPARTURE

CHAPTER I

THERE IS NOTHING OUTSIDE THE UNIVERSE

We humans are the species that makes things. So when we find something that appears to be beautifully and intricately structured, our almost instinctive response is to ask, 'Who made that?' The most important lesson to be learned if we are to prepare ourselves to approach the universe scientifically is that this is not the right question to ask. It is true that the universe is as beautiful as it is intricately structured. But it cannot have been made by anything that exists outside it, for by definition the universe is all there is, and there can be nothing outside it. And, by definition, neither can there have been anything before the universe that caused it, for if anything existed it must have been part of the universe. So the first principle of cosmology must be 'There is nothing outside the universe'.

This is not to exclude religion or mysticism, for there is always room for those sources of inspiration for those who seek them. But if it is knowledge we desire, if we wish to understand what the universe is and how it came to be that way, we need to seek answers to questions about the things we see when we look around us. And the answers can involve only things that exist in the universe.

This first principle means that we take the universe to be, by definition, a closed system. It means that the explanation for anything in the universe can involve only other things that also exist in the universe. This has very important consequences, each of which will be reflected many times in the pages that follow. One of the most important is that the *definition* or *description* of any entity inside the universe can refer only to other things in the universe. If something has a position, that position can be defined only with respect to the other things in the universe. If it has a motion, that motion can be discerned only by looking for changes in its position with respect to other things in the universe.

So, there is no meaning to space that is independent of the relationships among real things in the world. Space is not a stage, which might be either empty or full, onto which things come and go. Space is nothing apart from the things that exist; it is only an aspect of the relationships that hold between things. Space, then, is something like a sentence. It is absurd to talk of a sentence with no words in it. Each sentence has a grammatical structure that is defined by relationships that hold between the words in it, relationships like subject–object or adjective–noun. If we take out all the words we are not left with an empty sentence, we are left with nothing. Moreover, there are many different grammatical structures, catering for different arrangements of words and the various relationships between them. There is no such thing as an absolute sentence structure that holds for all sentences independent of their particular words and meanings.

The geometry of a universe is very like the grammatical structure of a sentence. Just as a sentence has no structure and no existence apart from the relationships between the words, space has no existence apart from the relationships that hold between the things in the universe. If you change a sentence by taking some words out, or changing their order, its grammatical structure changes. Similarly, the geometry of space changes when the things in the universe change their relationships to one another.

As we understand it now, it is simply absurd to speak of a universe with nothing in it. That is as absurd as a sentence with no words. It is even absurd to speak of a space with only one thing in it, for then there would be no relationships to define where that one thing is. (Here the analogy breaks down because there do exist sentences of one word only. However, they usually get their meaning from their relationships with adjacent sentences.)

The view of space as something that exists independent of any relationships is called the absolute view. It was Newton's view, but it has been definitively repudiated by the experiments that have verified Einstein's theory of general relativity. This has radical implications, which take a lot of thinking to get used to. There are unfortunately not a few good professional physicists who still think about the world as if space and time had an absolute meaning.

Of course, it does seem as though the geometry of space is not affected by things moving around. When I walk from one side of a room to the other, the geometry of the room does not seem to change. After I have crossed the room, the space within it still seems to satisfy the rules of Euclidean geometry that we learned in school, as it did before I started to move. Were Euclidean geometry not a good approximation to what we see around us, Newton would not have had a chance. But the apparent Euclidean geometry of space turns out to be as much an illusion as the apparent flatness of the Earth. The Earth seems flat only when we can't see the horizon. Whenever we can see far enough, from an aircraft or when we gaze out to sea, we can easily see that this is mistaken. Similarly, the geometry of the room you are in seems to satisfy the rules of Euclidean geometry only because the departures from those rules are very small. But if you could make very precise measurements you would find that the angles of triangles in your room do not sum to exactly 180 degrees. Moreover, the sum actually depends on the relation of the triangle to the stuff in the room. If you could measure precisely enough you would see that the geometries of all the triangles in the room do change when you move from one side of it to the other.

It may be that each science has one main thing to teach humanity, to help us shape our story of who we are and what we are doing here. Biology's lesson is natural selection, as its exponents such as Richard Dawkins and Lynn Margulis have so eloquently taught us. I believe that the main lesson of relativity and quantum theory is that the world is nothing but an evolving network of relationships. I have not the eloquence to be the Dawkins or Margulis of relativity, but I do hope that after reading this book you will have come to understand that the relational picture of space and time has implications that are as radical as those of natural selection, not only for science but for our perspective on who we are and how we came to exist in this evolving universe of relations.

Charles Darwin's theory tells us that our existence was not inevitable, that there is no eternal order to the universe that necessarily brought us into being. We are the result of processes much more complicated and unpredictable than the small aspects of our lives and societies over which we have some control. The lesson that the world is at root a network of evolving relationships tells us that this is true to a lesser or greater extent of all things. There is no fixed, eternal frame to the universe to define what may or may not exist. There is nothing beyond the world except what we see, no background to it except its particular history.

This *relational* view of space has been around as an idea for a long time. Early in the eighteenth century, the philosopher Gottfried Wilhelm Leibniz argued strongly that Newton's physics was fatally flawed because it was based on a logically imperfect absolute view of space and time. Other philosophers and scientists, such as Ernst Mach, working in Vienna at the end of the nineteenth century, were its champions. Einstein's theory of general relativity is a direct descendent of these views.

A confusing aspect of this is that Einstein's theory of general relativity can consistently describe universes that contain no matter. This might lead one to believe that the theory is not relational, because there is space but there is no matter, and there are no relationships between the matter that serve to define space. But this is wrong. The mistake is in thinking that the relationships that define space must be between material particles. We have known since the middle of the nineteenth century that the world is not composed only of particles. A contrary view, which shaped twentieth-century physics, is that the world is also composed of fields.

Fields are quantities that vary continuously over space, such as electric and magnetic fields.

The electric field is often visualized as a network of lines of force surrounding the object generating the field, as shown in Figure 1. What makes this a field is that there is a line of force passing through every point (as with a contour map, only lines at certain intervals are depicted). If we were to put a charged particle at any point in the field, it would experience a force pushing it along the field line that goes through that point.

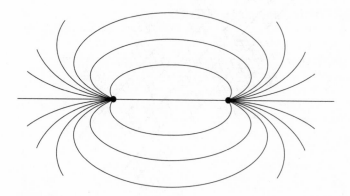

FIGURE I

The electric field lines between a positively and a negatively charged electron.

General relativity is a theory of fields. The field involved is called the gravitational field. It is more complicated than the electric field, and is visualized as a more complicated set of field lines. It requires three sets of lines, as shown in Figure 2. We may imagine them in different colours, say red, blue and green. Because there are three sets of field lines, the gravitational field defines a network of relationships having to do with how the three sets of lines link with one another. These relationships are described in terms of, for example, how many times one of the three kinds of line knot around those of another kind.

In fact, these relationships are all there is to the gravitational field. Two sets of field lines that link and knot in the same way define the same set of relationships, and exactly the same physical situation (an example is shown in Figure 3). This is why we call general relativity a relational theory.

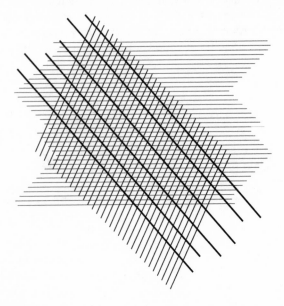

FIGURE 2
The gravitational field is like the electric field but requires three sets of field lines to describe it.

Points of space have no existence in themselves – the only meaning a point can have is as a name we give to a particular feature in the network of relationships between the three sets of field lines.

This is one of the important differences between general relativity and other theories such as electromagnetism. In the theory of electric fields it is assumed that points have meaning. It makes sense to ask in which direction the field lines pass at a given point. Consequently, two sets of electric field lines that differ only in that one is moved a metre to the left, as

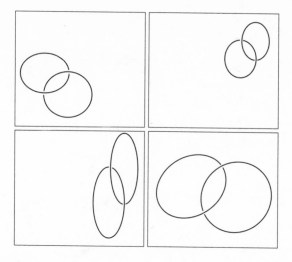

FIGURE 3

In a relational theory all that matters is the relationships between the field lines. These four configurations are equivalent, as in each case the two curves link in the same way.

in Figure 4, are taken to describe different physical situations. Physicists using general relativity must work in the opposite way. They cannot speak of a point, except by naming some features of the field lines that will uniquely distinguish that point. All talk in general relativity is about relationships among the field lines.

One might ask why we do not just fix the network of field lines, and define everything with respect to them. The reason is that the network of relationships evolves in time. Except for a small number of idealized examples which have nothing to do with the real world, in all the worlds that general relativity describes the networks of field lines are constantly changing.

This is enough for the moment about space. Let us turn now to time. There the same lesson holds. In Newton's theory time is assumed to have an absolute meaning. It flows, from the infinite past to the infinite future, the same everywhere in the universe, without any relation to things that actually happen. Change is measured in units of time, but time is assumed to

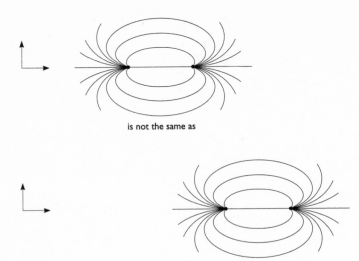

is not the same as

FIGURE 4
In a non-relational theory it matters also where the field lines are in absolute space.

have a meaning and existence that transcends any particular process of change in the universe.

In the twentieth century we learned that this view of time is as incorrect as Newton's view of absolute space. We now know that time also has no absolute meaning. There is no time apart from change. There is no such thing as a clock outside the network of changing relationships. So one cannot ask a question such as how fast, in an absolute sense, something is changing: one can only compare how fast one thing is happening with the rate of some other process. Time is described only in terms of change in the network of relationships that describes space.

This means that it is absurd in general relativity to speak of a universe in which nothing happens. Time is nothing but a measure of change – it has no other meaning. Neither space nor time has any existence outside the system of evolving relationships that comprises the universe. Physicists refer to this feature of general relativity as *background independence*.

By this we mean that there is no fixed background, or stage, that remains fixed for all time. In contrast, a theory such as Newtonian mechanics or electromagnetism is *background dependent* because it assumes that there exists a fixed, unchanging background that provides the ultimate answer to all questions about where and when.

One reason why it has taken so long to construct a quantum theory of gravity is that all previous quantum theories were background dependent. It proved rather challenging to construct a background independent quantum theory, in which the mathematical structure of the quantum theory made no mention of points, except when identified through networks of relationships. The problem of how to construct a quantum theoretic description of a world in which space and time are nothing but networks of relationships was solved over the last 15 years of the twentieth century. The theory that resulted is loop quantum gravity, which is one of our three roads. I shall describe what it has taught us in Chapter 10. Before we get there, we shall have to explore other implications of the principle that there is nothing outside the universe.

CHAPTER 2

....

IN THE FUTURE WE SHALL KNOW MORE

One of the things that cannot exist outside the universe is ourselves. This is obviously true, but let us consider the consequences. In science we are used to the idea that the observers must remove themselves from the system they study, otherwise they are part of it and cannot have a completely objective point of view. Also, their actions and the choices they make are likely to affect the system itself, which means that their presence may contaminate their understanding of the system.

For this reason we try as often as we can to study systems in which a clean boundary can be drawn separating the system under study from the observer. That we can do this in physics and astronomy is one of the reasons why those sciences are said to be 'harder'. They are held to be more objective and more reliable than the social sciences because in physics and astronomy there seems to be no difficulty with removing the observer from the system. In the 'softer' social sciences there is no way around the fact that the scientists themselves are participants in the societies they study. Of course, it is possible to try to minimize the effects of this and, for better or worse, much of the methodology of the social sciences is based on the belief that the more one can remove the observer from the system, the more scientific one is being.

This is all well and good when the system in question can be isolated, say in a vacuum chamber or a test tube. But what if the system we want to understand is the whole universe?

We do live in the universe, so we need to ask whether the fact that cosmologists are part of the system they are studying is going to cause problems. It turns out that it does, and this leads to what is probably the most challenging and confusing aspect of the quantum theory of gravity.

Actually, part of the problem has nothing to do with quantum theory, but comes from putting together two of the most important discoveries of the early twentieth century. The first is that nothing can travel faster than light; the second is that the universe seems to have been created a finite time ago. Current estimates put this time at about 14 billion years, but the exact number is not important. Together, the two things mean that we cannot see the whole universe. We can see only the contents of a region that extends around us to about 14 billion light years – the distance light could travel in this time. This means that science cannot, in principle, provide the answer to any question we might ask. There is no way to find out, for example, how many cats there are in the universe, or even how many galaxies there are. The problem is very simple: no observer inside the universe can see all of what is in the universe. We on Earth cannot receive light from any galaxy, or any cat, more than 14 billion or so light years from us. So if someone asserts that there are exactly 212,400,000,043 more cats in the universe than can be seen from Earth, no investigation we can do can prove them right or wrong.

However, the universe is quite likely to be much larger than 14 billion light years across. Why this is so would take us too far afield, but let me say simply that we have yet to find any evidence of the universe either ending or closing in on itself. There is no feature in what we can see that suggests that it is not just a small fraction of what exists. But if this is so, then even with perfect telescopes we would be able to see only a small part of all that exists.

Since the time of Aristotle, mathematicians and philosophers have investigated the subject of logic. Their aim has been to establish the laws by which we reason. And ever since its beginnings, logic has assumed that every statement is either true or false. Once this is assumed, it is possible to deduce true statements from other true statements. Unfortunately, this kind of logic is completely inapplicable when it comes to making deductions about the whole universe. Suppose we count all the cats in the region of the universe that we can see, and the number comes to one trillion. This is a statement whose truth we can establish. But what of a statement such as, 'Fourteen billion years after the big bang, there are a hundred trillion cats in the whole universe'? This may be true or false, but we observers on Earth have absolutely no way of determining which. There may be no cats farther than 14 billion light years from us, there may be 99 trillion or there may be an infinite number. Although these are all assertions that we can state, we cannot decide whether they are true or false. Nor can any other observer establish the truth of any claim as to the number of cats in the universe. Since it takes only about four billion years for cats to evolve on a planet, no observer could know whether cats have evolved in some region of space so far away from her that light reflected from their mysterious eyes could not have yet reached her.

However, classical logic demands that every statement be either true or false. Classical logic is therefore not a description of how we reason. Classical logic could be applied only by some being outside the universe, a being who could see the whole cosmos and count all its cats. But, if we insist on our principle that there is nothing outside the universe, there is no such being. To do cosmology, then, we need a different form of logic – one that does not assume that every statement can be judged true or false. In this kind of logic, the statements an observer can make about the universe are divided into at least three groups: those that we can judge to be true, those that we can judge to be false and those whose truth we cannot decide upon at the present time.

According to the classical view of logic, the question of whether a statement can be judged to be true or false is something absolute – it depends only on the statement and not on the observer doing the judging. But it is easy to see that this is not true in our universe, and the reason is closely related to what we have just said. Not only can an individual observer only see light from one part of the universe; which part they can see depends on where they find themselves in the history of the universe. We can judge the truth or falseness of statements about the Spice Girls. But observers who live more than 14 billion years from us cannot because they will not have received any information that would even let them suspect the existence of such a phenomenon. So we must conclude that the ability to judge whether a statement is true or false depends to some extent on the relationship between the observer and the subject of the statement.

Furthermore, an observer who lives on Earth a billion years from now will be able to see much more of the universe, for they will be able to see 15 billion light years out into the universe rather than the 14 billion light years we can see. They will see everything we can see, but they will see much more because they will see farther (Figure 5). They may be

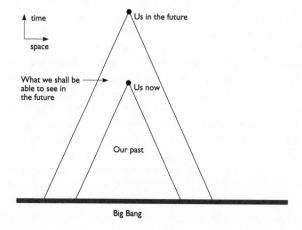

FIGURE 5

Observers in the future will be able to see more of the universe than we can see now. The diagonal lines represent the paths of light rays travelling from the past to us. Since nothing can travel faster than light, anything in our past that we can see or experience any effect of must lie within the triangle completed by the two diagonal lines. In the future we shall be able to receive light from farther away, and therefore see farther.

able to see many more cats. So, the list of statements they can judge to be true or false includes all that we can judge, but it is longer. Or consider an observer who lives 14 billion years after the big bang, as we do, but is 100 billion light years from us. Many cosmologists argue that the universe is at least 100 billion light years across; if they are right there is no reason for there not to be intelligent observers at that distance from us. But the part of the universe that they see has no overlap with the part of the universe that we see. The list of statements they can judge to be true or false is thus completely different from the list of statements that we here on Earth can judge to be true or false. If there is a logic that applies to cosmology, it must therefore be constructed so that which statements can be judged to be true or false depends on the observer. Unlike classical logic, which assumes that all observers can decide the truth or falsity of all statements, this logic must be observer-dependent.

In the history of physics it has often happened that by the time the physicists have been able to understand the need for a new mathematics, they found that the mathematicians had got there first and had already invented it. This is what happened with the mathematics needed for quantum theory and relativity and it has happened here as well. For reasons of their own, during the twentieth century mathematicians investigated a whole collection of alternatives to the standard logic we learned in school. Among them is a form of logic which we may call 'logic for the working cosmologist', for it incorporates all the features we have just described. It acknowledges the fact that reasoning about the world is done by observers inside the world, each of whom has limited and partial information about the world, gained from what they can observe by looking around them. The result is that statements can be not only true or false; they can also carry labels such as 'we can't tell now whether it's true, but we might be able to in the future'. This cosmological logic is also intrinsically observer-dependent, for it acknowledges that each observer in the world sees a different part of it.

The mathematicians, it seems, were not aware that they were inventing the right form of logic for cosmology, so they called it other names. In its first forms it was called 'intuitionistic logic'. More sophisticated versions which have been studied more recently are known collectively as 'topos theory'. As a mathematical formalism, topos theory is not easy. It is perhaps the hardest mathematical subject I've yet encountered. All of what I know of it comes from Fotini Markopoulou-Kalamara, who discovered that cosmology requires non-standard logic and found that topos theory was right for it. But the basic themes of it are obvious, for they describe our real situation in the world, and not only as cosmologists. Here in the real world, we almost always reason with incomplete information. Each day we encounter statements whose truth or falsity cannot be decided on the basis of what we know. And in the forms of our social and political life we recognize, often explicitly, that different observers have access to different information. We also deal every day with the fact that the truth or falsity of statements about the future may be affected by what we choose to do.

This has very profound implications for a whole host of issues. It means that to judge the rationality of our decisions, we do not have to pretend that there is some supernatural observer who knows everything: it is enough to demand that the different observers report what they see honestly. When this rule is followed we discover that when we and another person each have enough information to decide whether something is true or false, we always make the same decision.

Thus, the philosophers who attempted to ground ethics and science in the ultimate judgements of an all-knowing being were mistaken. We can live rationally without having to believe in a being who sees everything. We need only believe in the ethical principle that observers should communicate honestly what they see. If we stick to this, then the fact that there will always be questions that we cannot answer need not prevent us from coming to an agreement about how to understand those aspects of our world which we share in common.

So topos, or cosmological, logic is also the right logic for understanding the human world. It, and not Aristotle, must be the right basis for economics, sociology and political science. I am not aware that anyone in these areas has taken up topos theory and tried to make it the foundation of their subject, although George Soros's approach to economics, which he calls the theory of reflexivity, is certainly a start in the right direction. But we should not be surprised if both cosmology and social theory point us in the same direction. They are the two sciences that cannot be formulated sensibly unless we build into their foundations the simple fact that all possible observers are inside the systems they study.

CHAPTER 3

....

MANY OBSERVERS, NOT MANY WORLDS

So far I have said nothing at all about quantum theory. We
have seen that even without it, doing cosmology requires a
radical revision of our way of doing science – a revision that
goes even to the foundations of logic. Any scientific form of
cosmology requires a radical change in the logic we use, to
take into account the fact that the observer is inside the
universe. This requires us to build our theory so that from the
beginning it takes into account a form of observer dependence. We must acknowledge that each observer can have
only a limited amount of information about the world, and
that different observers will have access to different information.

With this important principle in mind, we may turn to the
problem of how to bring quantum theory into cosmology.
'Hold it!', I can hear the reader saying. 'Quantum theory is
confusing enough. Now I'm being asked to think about how to
apply it to the universe as a whole! Where do I get off?' That's
understandable, but, as I shall explain in this chapter,
thinking about how to apply quantum theory to the universe
as a whole may make quantum physics easier, not harder, to
understand. The principles we have looked at in the first two
chapters may very well be the key to making quantum theory
comprehensible.

Quantum theory is puzzling because it challenges our
standard ideas about the relationship between theory and
observer. The theory is indeed so puzzling that there is no

universally accepted physical interpretation of it. There are many different points of view about what quantum theory really asserts about reality and its relationship to the observer. The founders of quantum theory, such as Einstein, Bohr, Heisenberg and Schrödinger, could not agree on these questions. Nor is the present-day situation any better, for now we have extra points of view that those guys, smart as they were, were not imaginative enough to foresee. There is now no more agreement about what quantum theory means than when Einstein and Bohr first debated the question in the 1920s.

It is true that there is only one mathematical formalism for the quantum theory. So physicists have no problem with going ahead and using the theory, even though they do not agree about what it means. This may seem strange, but it does happen. I have worked on projects in quantum gravity where everything went smoothly until the collaborators discovered one day over dinner that we had radically different understandings of the meaning of quantum theory. Everything went smoothly again after we had calmed down and realized that how we thought about the theory had no effect on the calculations we were doing.

But this is no consolation to the layperson, who does not have the mathematics to fall back on. With only the concepts and principles to go on, it must be very disconcerting to discover that different physicists, in their different books, offer very different versions of the basics of quantum theory.

Quantum cosmology helps rather than hinders because, as we are about to see, it limits the scope for possible interpretations of the quantum theory. If we stick to the principles introduced in the first two chapters, several of the approaches to the interpretation of quantum mechanics must be abandoned. Either that, or we must give up any idea that quantum theory can be applied to space and time. The principle that there is nothing outside the universe and the principle that in the future we shall know more do point to a new way of looking at quantum theory that is both simpler and more rational than many of the older ideas. As a result of applying quantum theory to cosmology, there has emerged over the last few years a new approach to the problem of the meaning of quantum theory. This is what I want to communicate in this chapter.

Ordinary quantum theory is a theory of atoms and molecules. In the form developed originally by Bohr and Heisenberg, it required the world to be split into two parts. In one part was the system under study, which was described using the quantum theory, and in the other part lived the observer, together with whatever measuring instruments were needed to study the first system. This separation of the world into two parts is essential for the very structure of quantum mechanics. At the heart of this structure lies the *superposition principle*, which is one of the basic axioms of the quantum theory.

The superposition principle is not easy to understand, because it is formulated in seemingly abstract terms. If one is not careful it can lead to a kind of mysticism in which its meaning is over-interpreted far past what the evidence calls for. So we shall be careful, and spend some time looking at the statement of this important principle.

Let us first state it. The superposition principle says that if a quantum system can be found in one of two *states*, A and B, with different properties, it may also be found in a combination of them, $aA + bB$, where a and b are any numbers. Each such combination is called a superposition, and each is physically different.

But what does this actually mean? Let us break it down. The first thing to understand is what physicists mean when they talk about 'states'. This one word contains almost the full mystery of the quantum theory. Roughly, we say that the state of a physical system is its configuration at a particular moment. For example, if the system is the air in the room, its state might consist of the positions of all the molecules together with their speeds and the directions of their motions. If the system is a stock market, the state is the list of the prices of all the stocks at a particular moment. One way to say this is that a state consists of all the information needed to completely describe a system at an instant of time.

However, there is a problem with using this idea in quantum theory, because we are not able to measure at the same time both the position and the motion of a particle. Heisenberg's uncertainty principle asserts that we can only ever measure accurately *either* the position *or* the direction and speed of motion of a particle. For the moment, don't worry about why this should be. It is part of the mystery – and to be honest, no one really knows how it comes about. But let us look at its consequences.

If we cannot determine both the position and the motion of a particle, then the above definition of 'state' is no use to us. There may or may not be something in reality corresponding to the exact state, which comprises both the position and the motion, but, according to the uncertainty principle, even if it exists in some ideal sense it would not be a quantity we could observe. So in quantum theory we modify the concept of a state so that it refers only to as complete a description as may be given, subject to the restriction coming from the uncertainty principle. Since we cannot measure both the position and the motion, the possible states of the system can involve either a description of its exact position, or of its exact motion, but not both.

Perhaps this seems a bit abstract. It may also be hard to think about, because the mind rebels: it is hard to work one's way through to the logical consequences of a principle like the uncertainty principle when one's first response is simply to disbelieve it. I myself do not really believe it, and I do not think I am the only physicist who feels this way. But I persist in using it because it is a necessary part of the only theory I know that explains the main observed facts about atoms, molecules and the elementary particles.

So, if I want to speak about atoms without contradicting the uncertainty principle, I must conceive of states as being described by only some of the information I might be seeking. This is the first hard thing about states. As a state contains only part of the information about a system, there must be some rationale for that information being selected. However, although the uncertainty principle limits how much information a state can have, it does not tell us how it is decided which information to include and which to leave out.

There can be several reasons for this choice. It can have to do with the history of the system. It can have to do with the context the system now finds itself in, for example with how it is connected to, or correlated with, other things in the universe. Or it can have to do with a choice we, the observer, have made. If we choose to measure different quantities, or even in some circumstances to ask different questions, this can have an effect on the state. In all these cases the state of a system is not just a property of that system at a given time, but involves some element outside the present system, having to do either with its past or with its present context.

We are now ready to talk about the superposition principle. What could it possibly mean to say that if a system can be in state A or state B, it can also be in a combination of them, which we write as $aA + bB$, where a and b are numbers?

It is perhaps best to consider an example. Think of a mouse. From the point of view of a cat, there are two kinds of mice – tasty and yukky. The difference is a mystery to us, but you can be sure that any cat can tell them apart. The problem is that the only way to tell is to taste one. From the point of view of ordinary feline experience, any mouse is one or the other. But according to quantum theory this is a very coarse approximation to the way the world actually is. A real mouse, as opposed to the idealized version that Newtonian physics offers, will generally be in a state that is neither tasty nor yukky. It will instead have a probability that, if tasted, it will be one or the other – say, an 80 per cent chance of being tasty. This state of being suspended in between two states is not, according to quantum theory, anything to do with our influence – it really is neither one thing nor the other. The state may be anywhere along a whole continuum of possible situations, each of which is described by a quantum state. Such a quantum state is described by its having a certain propensity to be tasty and another propensity to be yukky; in other words, it is a superposition of two states – the states of purely tasty and purely yukky. This superimposed state is described mathematically by adding a certain amount of one to the other. The proportions of each are related to the probabilities that when bitten, the poor mouse will prove to be tasty or not.

This sounds crazy, and even thirty years after learning it I cannot describe this situation without a feeling of misgiving. Surely there must be a better way to understand what is going on here! Embarrassing though it is to admit it, no one has yet found a way to make sense of it that is both more comprehensible and elegant. (There are alternatives, but they are either comprehensible and inelegant, or the reverse.) However, there is a lot of experimental evidence for the superposition principle, including the double slit experiment and the Einstein–Podolsky–Rosen experiment. Interested readers can find these discussed in many popular books, some of which are included in the reading list at the end of this book.

The problem with quantum theory is that nothing in our experience behaves in the way the theory describes. All our perceptions are either of one thing or another – A or B, tasty or yukky. We never perceive combinations of them, such as $a \times$ tasty $+ b \times$ yukky. Quantum theory takes this into account. It says that what we observe will be tasty a certain proportion of the time, and yukky the rest of the time. The relative probabilities of us observing these two possibilities are given by the relative magnitudes of a^2 and b^2. However, what is most crucial to take on board is that the statement that the system is in the state aA + bB does not mean that it is either A or B, with some probability of being A and some other probability of being B. That is what we see if we observe it, but that is not what it is. We know this because the superposition aA + bB can have properties that neither tasty nor yukky have by themselves.

There is a paradox here. Were my cat to be described in the language of quantum theory, after tasting the mouse she would experience either tasty or yukky. But according to quantum mechanics she would not be in a definite state of happy or displeased. She would go into a superposition of two states which mirrors the possible states of the mouse. She would be suspended in a superposition of a happy state and an annoyed-for-having-bitten-into-a-yukky-mouse state.

So the cat experiences herself in a definite state, but in the light of quantum theory I must see her in a superposition.

Now, what happens if I observe my cat? I shall certainly experience a purr or a scratch. But shall I definitely be in one of these two possible states? I cannot imagine that I should not experience one or the other. I cannot imagine even what it would mean to experience anything other than one or the other. But if I am described in the language of quantum theory, I too, along with the mouse and the cat, will be in a superposition of two different states. In one of them the mouse was tasty, the cat was happy and I heard a purr. In the other the mouse is yukky, the cat is angry and I am nursing a scratch.

What makes the theory consistent is that our different states are correlated. My being happy goes along with the happiness of the cat and the tastiness of the mouse. If an observer queries both me and the cat, our answers will be consistent, and they will even be consistent with the observer's experience if she tastes the mouse. But none of us is in a definite state. According to quantum theory, we are all in a superposition of the two possible correlated states. The root of the apparent paradox is that my own experience is of one thing or the other, but the description of me that would be given in quantum theory by another observer has me most often in a superposition which is none of the things I actually experience.

There are a few possible resolutions of this mystery. One is that I am simply mistaken about the impossibility of superpositions of mental states. In fact, if the usual formalism of quantum mechanics is to be applied to me, as a physical system, this must be the case. But if a human being can be in a superposition of quantum states, should the same not be true of the planet Earth? The solar system? The Galaxy? In fact, why should it not be a physical possibility that the whole universe is in a superposition of quantum states? Since the 1960s there have been a series of efforts to treat the whole universe in the same way as we treat quantum states of atoms. In these descriptions of the universe in terms of quantum states, it is assumed that the universe may as easily be put into quantum superpositions as can states of photons and electrons. This subject can therefore be called 'conventional quantum cosmology' to distinguish it from other approaches to combining quantum theory and cosmology that we shall come to.

In my opinion, conventional quantum cosmology has not been a success. Perhaps this is too harsh a judgement. Several of the people I most respect in the field disagree with this. My own views on the matter have been shaped by experience as much as reflection. By chance I was part of the discovery of the first actual solutions to the equations that define a quantum theory of cosmology. These are called the Wheeler–DeWitt equations or the quantum constraints equations. The solutions to these equations define quantum states that are meant to describe the whole universe.

Working first with one friend, Ted Jacobson, then with another, Carlo Rovelli, I found an infinite number of solutions to these equations in the late 1980s. This was very surprising, as very few of the equations of theoretical physics can be solved exactly. One day in February 1986, Ted and I, working in Santa Barbara, set out to find approximate solutions to the equations of quantum cosmology, which we had been able to simplify thanks to some beautiful results obtained by two friends, Amitaba Sen and Abhay Ashtekar. All of a sudden we realized that our second or third guess, which we had written on the blackboard in front of us, solved the equations exactly. We tried to compute a term that would measure how much our results were in error, but there was no error term. At first we looked for our mistake, then all of a sudden we saw that the expression we had written on the blackboard was spot on: an exact solution of the full equations of quantum gravity. I still remember vividly the blackboard, and that it was sunny and Ted was wearing a T-shirt (then again, it is always sunny in Santa Barbara and Ted always wears a T-shirt). This was the first step of a journey that took ten years, sometimes exhilarating and often aggravating years, before we understood what we had really found in those few minutes.

Among the things we had to struggle with were the implications of the fact that the observer in quantum cosmology is inside the universe. The problem is that in all the usual interpretations of quantum theory the observer is assumed to be outside the system. That cannot be so in cosmology. This is our principle and, as I've emphasized before, this is the whole point. If we do not take it into account, whatever we may do is not relevant to a real theory of cosmology.

Several different proposals for making sense of the quantum theory of the whole universe had been put forward by pioneers of the subject such as Francis Everett and Charles Misner. We were certainly aware of them. For many years young theoretical physicists have amused themselves by debating the merits and absurdities of the different proposals made for quantum cosmology. At first this feels fantastic – one is wrestling with the very foundations of science. I used to look at the older people and wonder why they never seemed to spend their time this way. After a while I understood: one could only go around the five or six possible positions a few dozen times before the game got very boring. Something was missing.

So we did not exactly relish the idea of taking on this problem. Indeed, at least for me, solving equations rather than worrying about foundations was a deliberate strategy to try to do something that could lead to real progress. I had spent much of my college years staring at the corner of my room, wondering about what was real in the quantum world. That was good for then; now I wanted to do something more positive. But this was different, for in a flash we had obtained an infinite number of absolutely genuine solutions to the real equations of quantum gravity. And if a few were very simple, most were exceedingly complex – as complex as the most complicated knot one could imagine (for they indeed had something to do with tying knots, but we shall come to that later on).

No one had ever had to, or been able to, contemplate the meaning of these equations in anything other than very drastic approximations. In these approximations the complexity and wonder of the universe is cut down to one or two variables, such as how big the universe is and how fast it is expanding. It is very easy to forget one's place and fall for the fantasy that one is outside the universe, having reduced the history of the universe to a game as simple as playing with a yo-yo. (No, actually simpler, for we never would have been able to attack something as complicated as a real yo-yo. The equations we used to model what we optimistically called 'quantum cosmology' were something like a description of a really stupid yo-yo, one that can only go up and down, never forward or back or to the left or right.)

What is needed is an interpretation of the states of quantum theory that allows the observer to be part of the quantum system. One of the ideas on the table was presented by Hugh Everett in his hugely influential Ph.D. thesis of 1957. He invented a method called the relative state interpretation which allows you to do something very interesting. If you know exactly what question you want to ask, and can express it in the language of the quantum theory, then you can deduce the probabilities of different answers, even if the measuring instruments are part of the quantum system. This is a step forward, but we have still not really eliminated the special role that observations have in the theory. In particular this applies equally to an infinite set of questions that may be asked, all of which are mathematically equivalent from the point of view of the theory. There is nothing in the theory that tells us why the observations we make, in terms of big objects that appear to have definite positions and motions, are special. There is nothing to distinguish the world we experience from an infinite number of other worlds made up of complicated superpositions of things in our world.

We are used to the idea that a physical theory can describe an infinitude of different worlds. This is because there is a lot of freedom in their application. Newton's physics gives us the laws by which particles move and interact with one another, but it does not otherwise specify the configurations of the particles. Given any arrangement of the particles that make up the universe, and any choices for their initial motions, Newton's laws can be used to predict the future. They thus apply to any possible universe made up of particles that move according to their laws. Newton's theory describes an infinite number of different worlds, each connected with a different solution to the theory, which is arrived at by starting with the particles in different positions. However, each solution to Newton's theory describes a single universe. This is very different from what seems to be coming out of the equations of the conventional approach to quantum cosmology. There, each solution seems to have within it descriptions of an infinite number of universes. These universes differ, not only in the answers that the theory gives to questions, but by the questions that are asked.

Everett's relative-state form of the theory must therefore be supplemented by a theory of why what we observe corresponds to the answers to certain questions, and not to an infinite number of other questions. Several people have attempted to deal with this, and some progress has been made using an idea called *decoherence*. A set of questions is called decoherent if there is no chance that a definite answer to one is a superposition of definite answers to others. This idea has been developed by several people into an approach to quantum cosmology called *the consistent histories formulation*. This approach lets you specify a series of questions about the history of the universe. Assuming only that the questions are consistent with one another, in the sense that the answer to one will not preclude our asking another, this approach tells us how to compute the probabilities of the different possible answers. This is progress, but it does not go far enough. The world we experience is decoherent but, as has been convincingly shown by two young English physicists, Fay Dowker and Adrian Kent, so are an infinite number of other possible worlds.

One of the most dramatic moments I've experienced during my career in science was the presentation of this work at a conference on quantum gravity in Durham, England, in the summer of 1995. When Fay Dowker began her presentation on the consistent histories formulation, that approach was generally regarded as the best hope for resolving the problems of quantum cosmology. She was a postdoc under James Hartle, who had pioneered the development of the consistent histories approach to quantum cosmology, and there was little indication in her introduction of what was coming. In a masterful presentation she built up the theory, elucidating along the way some of its most puzzling aspects. The theory seemed in better shape than ever. Then she proceeded to demonstrate two theorems that showed that the interpretation did not say what we thought it did. While the 'classical' world we observe, in which particles have definite positions, may be one of the consistent worlds described by a solution to the theory, Dowker and Kent's results showed that there had to be an infinite number of other worlds too. Moreover, there were an infinite number of consistent worlds that have been classical up to this point but will not be anything like our world in five minutes' time. Even more disturbing, there were worlds that were classical now that were arbitrarily mixed up superpositions of classical at any point in the past. Dowker concluded that, if the consistent-histories interpretation is correct, we have no right to deduce from the existence of fossils now that dinosaurs roamed the planet a hundred million years ago.

I cannot speak for everyone who was in that room, but the people sitting near me were as shocked as I was. In conversations we had later that summer, Jim Hartle insisted that the work he and Murray Gell-Mann had done on the consistent histories approach was not contradicted by anything Fay Dowker had said. They were fully aware that their proposal imposed on reality a radical context dependence: one cannot talk meaningfully about the existence of any object or the truth of any statement without first completely specifying the questions that are to be asked. It is almost as if the questions bring reality into being. If one does not first ask for a history of the world that includes the question of whether dinosaurs roamed the Earth a hundred million years ago, one may not get a description in which the notion of dinosaurs – or any other big 'classical objects' – has any meaning.

I checked, and Hartle was right. What he and Gell-Mann had said was still valid. An interesting thing seems to have happened, which in retrospect is not all that unusual: many of us working on this problem had misunderstood Gell-Mann and Hartle to mean something much less radical, and much more comfortable to our classical, old-fashioned intuitions, than what they had actually proposed. There is, according to them, one history of the world, and it is expressed in quantum language. But this one world contains many different, equally consistent histories, each of which can be brought into being by the right set of questions. Each history is incompatible with the others, in the sense that they cannot be experienced together by observers like ourselves. But each is, according to the formalism, equally real.

As you might imagine, there was a huge, if mostly friendly, disagreement over what to make of this. Some of us follow Fay Dowker and Adrian Kent in their conviction that this infinite expansion of the notion of reality is unacceptable. Either quantum mechanics is wrong when applied to the whole universe, or it is incomplete in that it must be supplemented by a theory of which set of questions corresponds to reality. Others follow James Hartle and Murray Gell-Mann in embracing the extreme context dependence that comes with their formulation. As Chris Isham says, the problem lies with the meaning of the word 'is'.

If this were not trouble enough, there are other difficulties with this conventional formulation of quantum cosmology. It turns out that one is not free to ask any set of questions: instead, these are constrained by having to be solutions to certain equations. And, although we had solved the equations that determine the quantum states of the universe, it proved much more difficult to determine the questions that can be asked of the theory. It seems unlikely that this can ever be done – at least in any real theory, as opposed to the toy models that describe little yo-yo-like versions of the universe. Perhaps I should not comment on the likelihood of finding the right set of questions, given that our solving the equations for the states was itself a total accident. Still, we have tried, many smarter people have tried, and we have all come to the conclusion this is not a stone that can be moved. So conventional quantum cosmology seems to be a theory in which we can formulate the answers, but not the questions.

Of course, from the perspective of the last chapter, this is not surprising. We saw there that to formulate a theory of cosmology we must acknowledge that different observers see partly different, partial views of the universe. From this starting point it makes no sense to try to treat the whole universe as it if were a quantum system in a laboratory of the kind that ordinary quantum theory applies to. Could there be a different kind of quantum theory, one in which the quantum states refer explicitly to the domain seen by some observer? Such a theory would be different from conventional quantum theory. It would in a sense 'relativize' that theory, in the sense that it would make the quantum theory depend more explicitly on the location of the observer inside the universe. It would describe a large, perhaps infinite set of quantum worlds, each of which corresponds to the part of the world that could be seen by a particular observer, at a particular place and time in the history of the universe.

In the past few years there have been several proposals for just such a new kind of quantum cosmology. One of them grew out of the consistent-histories approach. It is a kind of reformulation of it, by Chris Isham and his collaborator Jeremy Butterfield, in which they make context dependence the central feature of the mathematical formulation of the theory. They found that they can do this using topos theory, which allows one to describe many interrelated quantum mechanical descriptions, differing according to choice of context, in one mathematical formalism. Their work is beautiful, but difficult in the way a philosopher like Hegel or Heidegger is difficult. It is not easy to find the right language to use to talk about the world if one really believes that the notion of reality depends on the context of the person doing to the talking.

For many of us in quantum gravity, Chris Isham is a kind of theorists' theorist. Most theoretical physicists think in terms of examples, and then seek to generalize what they have learned as widely as possible. Chris Isham seems to be one of the few people who can work productively in the other direction. Several times he has introduced important ideas in a very general form, leaving it to others to apply the lessons to particular examples. On one occasion this led to loop quantum gravity, when Carlo Rovelli saw in a very general idea of his a strategy that could pay off in very concrete terms. Something like this is happening now. People have been thinking about context dependence in quantum cosmology for about ten years. We have learned from Chris Isham what kind of mathematics we need to do this.

Before Isham and his collaborators, Louis Crane, Carlo Rovelli and I developed different versions of an idea we called relational quantum theory. Going back to our earlier feline example, the basic idea was that all the players have a context, which consists of the part of the world they describe. Rather than asking which quantum description is right – that of the mouse, the cat, me, my friend – we argued that one has to accept them all. There are many quantum theories, corresponding to the many different possible observers. They are all interrelated, because when two observers are able to ask the same question they must get the same answer. The mathematics of topos theory, as developed by Chris Isham and collaborators, has told us how to do this for any possible case in which it may arise.

A third context-dependent theory was formulated by Fotini Markopoulou-Kalamara, by extending her proposal for cosmological logic to quantum theory. The result is that a context turns out to be the past of an observer, at a given moment. This is a beautiful unification of quantum theory and relativity in which the geometry of light rays, that determines how information may travel, itself determines the possible contexts.

In all these theories there are many quantum descriptions of the same universe. Each of them depends on a way of splitting the universe into two parts such that one part contains the observer and the other part contains what the observer wishes to describe. Each such division gives a quantum description of part of the universe; each describes what one particular observer will see. All these descriptions are different, but they have to be consistent with one another. This resolves the paradox of superpositions by making it a consequence of one's point of view. The quantum description is always the description of some part of the universe by an observer who remains outside it. Any such quantum system can be in a superposition of states. If you observe a system that includes me, you may see me in a superposition of states. But I do not describe myself in such terms, because in this kind of theory no observer ever describes themself.

Many of us believe that this is a definite step in the right direction. Rather than trying to make sense of metaphysical statements about their being many universes – many realities – within one solution to the theory of quantum cosmology, we are constructing a pluralistic version of quantum cosmology in which there is one universe. That universe has, however, many different mathematical descriptions, each corresponding to what a different observer can see when they look around them. Each is incomplete, because no observer can see the whole universe. Each observer, for example, excludes themselves from the world they describe. But when two observers ask the same question, they must agree. And if I look around tomorrow it cannot happen that the past changed. If I see dinosaurs roaming today on a planet a hundred million light years away, they will still be roaming there when I receive signals from the planet next year.

Like all advocates of new ideas, we support our opinions with slogans as well as with results. Our slogans are 'In the future we shall know more' and 'One universe, seen by many observers, rather than many universes, seen by one mythical observer outside the universe'.

....

CHAPTER 4

THE UNIVERSE IS MADE OF PROCESSES, NOT THINGS

Imagine you are trying to explain to someone why you are so enamoured of your new girlfriend or boyfriend, and someone quite sensibly asks you to describe them. Why do our efforts on such occasions seem so inadequate? Your intuition tells you that there is something essential about this person, but it is very hard to put it into words. You describe what they do for a living, what they like to do for fun, what they look like, how they act, but somehow this does not seem to convey what they are really like.

Or imagine that you have fallen into one of those interminable discussions about culture and national characteristics. It seems so obvious that the English are different from the Greeks, who are nothing at all like the Italians, except that they are both different from the English in the same way. And how is it that the Chinese seem in certain ways a bit American in their spirit, when their cultural history is so different and so much older? Again, it seems that there is something real here, but most of our attempts to capture it in words seem to fall short of what we are trying to express.

There is a simple solution to these quandaries: tell a story. If we narrate the story of our new friend's life, where and how they grew up, who their parents are and how they raised them, where they studied, what happened in their past relationships, we communicate more of what is important about them than if we attempt to describe how they are now. The same goes for cultures. It is only when we know something about their histories, both recent and ancient, that we begin to gain any insight into why being human is expressed a bit differently in different parts of the world. This may be obvious, but why should it be so? What is it about a person or a culture that makes it so hard to describe without telling a story? The answer is that we are not dealing with a thing, like a rock or a can opener. These are objects which remain more or less the same from decade to decade. They can be described, for most purposes, as static objects, each with some collection of unchanging properties. But when we are dealing with a person or a culture we are dealing with a process that cannot be comprehended as a static object, independently of its history. How it is now is incomprehensible without knowing how it came to be.

Just what is it about a story that tells us so much? What extra information are we conveying when we tell a story? When we tell a story about someone we narrate a series of episodes in their life. These tell us something about that person because we believe, from having heard and understood many such stories, that what happens to a person as they grow up has an effect on who they are. We also believe that people's characters are best revealed in how they react to situations, both propitious and adverse, and in what they have sought to do or become.

However, it is not the events themselves that carry the information in a narration. A mere list of events is very boring and is not a story. This is perhaps what Andy Warhol was trying to convey in his movies of haircuts or of a day in the life of the Empire State Building. What makes a story a story is the connections between the events. These may be made explicit, but they often do not need to be, because we fill them in almost unconsciously. We can do that because we all believe that events in the past are to some extent the causes of events in the future. We can debate to what extent a person is shaped by what happens to them, but we do not need to be devout determinists to have a practical and almost instinctive understanding of the importance of causality. It is this understanding of causality that makes stories so useful. Who did what to whom, and when, and why, is interesting because of what we know about the consequences of actions and events.

Imagine what life would be like without causality. Suppose that the history of the world were no more than random sets of events with no causal connections at all between them. Things would just happen; nothing would remain in place. Furniture, houses, everything would just come into being and disappear. Can you imagine what that would really be like? I can't – it is far too different from the world we live in. It is causality that gives our world its structure, that explains why this morning our chairs and tables are in the same places we left them last night. And it is because of the overwhelming importance of causal relations in shaping our world that stories are much more informative than descriptions.

So it seems there are two kinds of thing in the world. There are objects like rocks and can openers that simply are, that may be explained completely by a list of their properties. And then there are things that can only be comprehended as processes, that can only be explained by telling stories. For things of this second kind a simple description never suffices. A story is the only adequate description of them because entities like people and cultures are not really things, they are processes unfolding in time.

Here is an idea for an art piece. Take a film which everyone has seen and loved, and extract from it a series of stills, one from every ten seconds of the film. Mount these in a large gallery, arranged sequentially. Invite people to view the film one still at a time. Would this be enjoyable? No, people might laugh a bit at the beginning, but most would quickly become bored. Of the few who looked at the whole film, many would be film-makers and critics who would be able to pick up some tricks about how the film was made. For most of the rest of us a film presented one still at a time would be quite uninteresting, even if it took no longer to view the whole sequence than to watch the film. Of course, when we watch a film we are really looking at a sequence of still images, presented to us at such a rate that we see movement. This is sometimes described by saying that the sequence of still images creates the illusion of motion, but that is not quite right. It is the still images themselves that are the illusion. The world is never still – it is always in motion. The illusion that photography creates is of a frozen moment of time. It corresponds to nothing in reality, nor is it itself real, for any photograph is also a process. In a few years it will fade as a result of chemical processes which are always going on between the molecules that make up the apparently still image. So what happens in a movie is that the real world of motion and change is recreated from a sequence of illusions, not the reverse.

We humans seem to be fascinated by our ability to hold back change for long periods of time. This may be why painting and sculpture are so fascinating and so valuable, for they offer the illusion of time stopped. But time cannot be stopped. A marble sculpture may look the same from day to day, but it is not: each day the surface becomes a little different as the marble interacts with the air. As the Florentines have learned only too well from the damage wrought to their heritage by pollution, marble is not an inert thing, it is a process. All the skill of the artist cannot turn a process into a thing, for there are no things, only processes that appear to change slowly on our human timescales. Even objects that seem not to change, like rocks and can openers, have stories. It is just that the timescale over which they change significantly is longer than for most other things. Geologists and cultural historians are very interested in narrating the stories of rocks and can openers.

So there are not really two categories of things in the world: objects and processes. There are only relatively fast processes and relatively slow processes. And whether it is a short story or a long story, the only kind of explanation of a process that is truly adequate is a story.

The illusion that the world consists of objects is behind many of the constructs of classical science. Supposing one wants to describe a particular elementary particle, say a proton. In the Newtonian mode of description one would describe what it is at a particular moment of time: where it is located in space, what its mass and electric charge are, and so forth. This is called describing the 'state' of the particle. Time is nowhere in this description; it is, indeed, an optional part of the Newtonian world. Once one has adequately described how something is, one then 'turns on' time and describes how it changes. To test a theory, one makes a series of measurements. Each measurement is supposed to reveal the state of the particle, frozen at some moment of time. A series of measurements is like a series of movie stills – they are all frozen moments.

The idea of a state in Newtonian physics shares with classical sculpture and painting the illusion of the frozen moment. This gives rise to the illusion that the world is composed of objects. If this were really the way the world is, then the primary description of something would be how it is, and change in it would be secondary. Change would be nothing but alterations in how something is. But relativity and quantum theory each tell us that this is not how the world is. They tell us – no, better, they scream at us – that our world is a history of processes. Motion and change are primary. Nothing *is*, except in a very approximate and temporary sense. How something is, or what its state is, is an illusion. It may be a useful illusion for some purposes, but if we want to think fundamentally we must not lose sight of the essential fact that 'is' is an illusion. So to speak the language of the new physics we must learn a vocabulary in which process is more important than, and prior to, stasis. Actually there is already available a suitable and very simple language which you will have no trouble understanding.

From this new point of view, the universe consists of a large number of *events*. An event may be thought of as the smallest part of a process, a smallest unit of change. But do not think of an event as a change happening to an otherwise static object. It is just a change, no more than that.

The universe of events is a *relational universe*. That is, all its properties are described in terms of relationships between the events. The most important relationship that two events can have is *causality*. This is the same notion of causality that we found was essential to make sense of stories. We say that an event, let us call it A, is in part the cause of another event, B, if A was necessary for B to occur. If A had not occurred, B could not have. In this case we can say that A was a *contributing cause* of the event B. An event may have more than one contributing cause, and an event may also contribute to causing more than one future event.

Given any two events, A and B, there are only three possibilities: either A is a cause of B, or B is a cause of A, or neither is the cause of the other. We say that in the first case A is in the *causal past* of B, in the second, B is in the causal past of A, and in the third case neither is in the causal past of the other. This is illustrated in Figure 6, in which each event is indicated by a point and each arrow represents a causal relation. Such a picture is a picture of the universe as a process. Figure 7 shows a more complicated universe, consisting of many events, with a complicated set of causal relationships. These pictures are stories told visually – diagrams of the history of a universe.

Such a universe has time built into it from the beginning. Time and change are not optional, for the universe is a story and it is composed of processes. In such a world, time and causality are synonymous. There is no meaning to the past of an event except the set of events that caused it. And there is no meaning to the future of an event except the set of events it will influence. When we are dealing with a causal universe, we can therefore shorten 'causal past' and 'causal future' to simply 'past' and 'future'. Figure 8 shows the causal past and

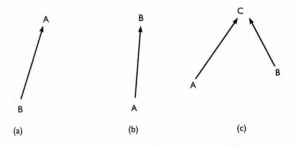

(a)   (b)   (c)

FIGURE 6

The three possible causal relations between two events, A and B: (a) A is to the future of B; (b) B is to the future of A; (c) A and B are neither to the future nor to the past of each other (though they may have other causal relations, for example both being in the past of event C, as shown).

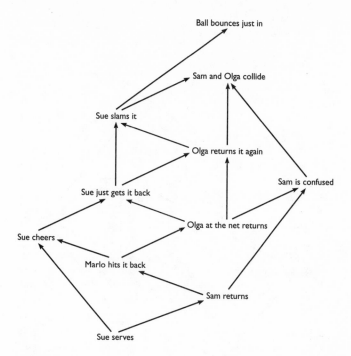

FIGURE 7
One volley in a tennis game, represented by the causal relations of a few of
its events.

future of a particular event in Figure 7. A causal universe is
not a series of stills following on, one after the other. There is
time, but there is not really any notion of a moment of time.
There are only process that follow one another by causal
necessity. It makes no sense to say what such a universe is. If
one wants to talk about it, one has no alternative but to tell its
story.

One way to think of such a causal universe is in terms of the
transfer of information. We can think of the content of each
arrow in Figures 6 to 8 as a few bits of information. Each event
is then something like a transistor that takes in information
from events in its past, makes a simple computation and
sends the result to the events in its future. A computation is

This applies to the origion of life.

Causal necessity demands more complexity in chemical reactions from the flow of solar energy

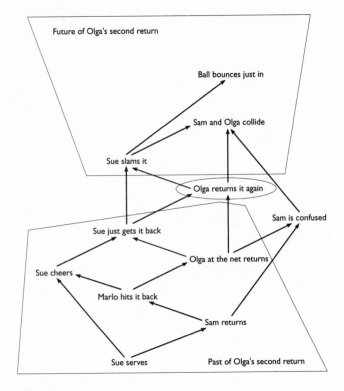

FIGURE 8
The future and past of Olga's second return. Note that Sam being confused is in neither set of events.

then a kind of story in which information comes in, is sent from transistor to transistor, and is occasionally sent to the output. If we were to remove the inputs and outputs from modern computers, most of them would continue to run indefinitely. The flow of information around the circuits of a computer constitutes a story in which events are computations and causal processes are just the flow of bits of information from one computation to the next. This leads to a very useful metaphor – the universe as a kind of computer. But it is a computer in which the circuitry is not fixed, but can evolve in time as a consequence of the information flowing through it.

Is our universe such a causal universe? General relativity tells us that it is. The description of the universe given by general relativity is exactly that of a causal universe, because of the basic lesson of relativity theory: that nothing can travel faster than light. In particular, no causal effect and no information can travel faster than light. Keep this in mind, and consider two events in the history of our universe, pictured in Figure 9. Let the first be the invention of rock and roll, which took place perhaps somewhere in Nashville in the 1950s. Let the second be the fall of the Berlin Wall, in 1989. Did the first causally influence the second? One may argue about the political and cultural influence of rock and roll, but what is important is only that the invention of rock and roll certainly had some effect on the events leading to the fall of the Berlin Wall. The people who first climbed the wall in triumph had rock and roll songs in their heads, and so did the functionaries who made the decisions that led to the reunification of Germany. So there was certainly a transfer of information from Nashville in the 1950s to Berlin in 1989.

FIGURE 9

The invention of rock and roll was in the causal past of the fall of the Berlin Wall because information was able to travel from the first event to the second.

So in our universe we define the causal future of some event to consist of all the events that it could send information to, using light or any other medium. Since nothing can travel faster than light, the paths of light rays leaving the event define the outer limits of the causal future of an event. They

Speed of light refers to photons not quarks!

form what we call the *future light cone* of an event (Figure 10). We call it a cone because, if we draw the picture so that space has only two dimensions, as in Figure 10, it looks like a cone. The causal past of an event consists of all the events that could have influenced it. The influence must travel from some event in the past at the speed of light or less. So the light rays arriving at an event form the outer boundary of the past of an event, and make up what we call the *past light cone* of an event. One is pictured in Figure 10. We can see that the structure of the causal relations around any event can be pictured in terms of both the past and future light cones. We

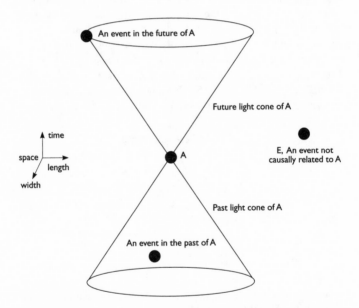

FIGURE 10

The past and future light cones of an event, A. The future light cone is made up of the paths of all light signals from A to any event in A's future. Any event inside the cone is in the future of A, causally, because an influence could travel from A to that event at less than the speed of light. We also see the past light cone of A, which contains all the events that may have influenced A. We also see another event, E, which is in neither the past nor the future of A. The diagram is drawn as if space had two dimensions.

see from Figure 10 also that there are many other events which lie outside both the past and future light cones of our particular event. These are events that took place so far from our event that light could not have reached it. For example, the birth of the worst poet in the universe, on a planet in a galaxy thirty billion light years from us is, fortunately, outside both our future and past light cones. So in our universe, specifying the paths of all the light rays or, equivalently, drawing the light cones around every event, is a way to describe the structure of all possible causal relations. Together, these relations comprise what we call the *causal structure* of a universe.

Many popular accounts of general relativity contain a lot of talk about 'the geometry of spacetime'. But actually most of that has to do with the causal structure. Almost all of the information needed to construct the geometry of spacetime consists of the story of the causal structure. So not only do we live in a causal universe, but most of the story of our universe is the story of the causal relations among its events. The metaphor in which space and time together have a geometry, called the spacetime geometry, is not actually very helpful in understanding the physical meaning of general relativity. That metaphor is based on a mathematical coincidence that is helpful only to those who know enough mathematics to make use of it. The fundamental idea in general relativity is that the causal structure of events can itself by influenced by those events. The causal structure is not fixed for all time. It is dynamical: it evolves, subject to laws. The laws that determine how the causal structure of the universe grows in time are called the *Einstein equations*. They are very complicated, but when there are big, slow moving klutzes of matter around, like stars and planets, they become much simpler. Basically, what happens then is that the light cones tilt towards the matter, as shown in Figure 11. (This is what is often described as the curvature, or distortion of the geometry of space and time.) As a result, matter tends to fall towards massive objects. This is, of course, another way of talking about the gravitational force. If matter moves around, then waves travel through the causal structure and the light cones

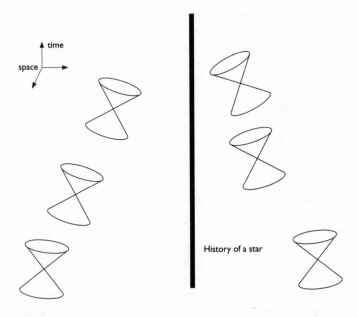

FIGURE 11

A massive object such as a star causes the light cones in its vicinity to tip towards it. This has the effect of causing freely falling particles to appear to accelerate towards the object.

oscillate back and forth, as shown in Figure 12. These are the *gravitational waves*.

So, Einstein's theory of gravity is a theory of causal structure. It tells us that the essence of spacetime is causal structure and that the motion of matter is a consequence of alterations in the network of causal relations. What is left out from the notion of causal structure is any measure of quantity or scale. How many events are contained in the passage of a signal from you to me, when we talk on the telephone? How many events have there been in the whole history of the universe in the past of this particular moment, as you finish reading this sentence? If we knew the answers to these questions, and we also knew the structure of causal relations among the events in the history of the universe, then we

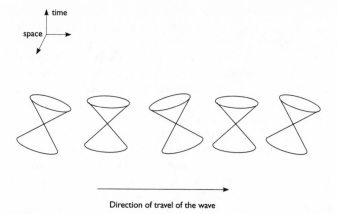

FIGURE 12

A gravitational wave is an oscillation in the directions in which the light cones point in spacetime. Gravitational waves travel at the speed of light.

would know all of what there is to know about the history of the universe.

There are two kinds of answer we could give to the question of how many events there are in a particular process. One kind of answer assumes that space and time are continuous. In this case time can be divided arbitrarily finely, and there is no smallest possible unit of time. No matter what we think of, say the passage of an electron across an atom, we can think of things that happen a hundred times faster. Newtonian physics assumes that space and time are continuous. But the world is not necessarily like that. The other possibility is that time comes in discrete bits, which can be counted. The answer to the question of how many events are required to transfer a bit of information over a telephone line will then be a finite number. It may be a very large number, but it still will be a finite number. But if space and time consist of events, and the events are discrete entities that can be counted, then space and time themselves are not continuous. If this is true, one cannot divide time indefinitely. Eventually we shall come to the elementary events, ones which cannot be further divided and are thus the simplest possible things that can happen.

Just as matter is composed of atoms, which can be counted, the history of the universe is constructed from a huge number of elementary events.

What we already know about quantum gravity suggests that the second possibility is right. The apparent smoothness of space and time are illusions; behind them is a world composed of discrete sets of events, which can be counted. Different approaches give us different pieces of evidence for this conclusion, but they all agree that if we looks finely enough at our world the continuity of space and time will dissolve as surely as the smoothness of material gives way to the discrete world of molecules and atoms.

The different approaches also agree about how far down we have to probe the world before we come to the elementary events. The scales of time and distance on which the discrete structure of the world becomes manifest is called the *Planck scale*. It is defined as the scale at which the effects of gravity and quantum phenomena will be equally important. For larger things, we can happily forget about quantum theory and relativity. But when we get down to the Planck scale we have no choice but to take it all into account. To describe the universe at this scale we need the quantum theory of gravity.

The Planck scale can be established in terms of known fundamental principles. It is calculated by putting together in appropriate combinations the constants that come into the fundamental laws. These are Planck's constant, from quantum theory; the speed of light, from special relativity; and the gravitational constant, from Newton's law of gravitation. In terms of the Planck scale, we are absolutely huge. The Planck length is 10^{-33} centimetres, which is 20 orders of magnitude smaller than an atomic nucleus. On the scale of the fundamental time, everything we experience is incredibly slow. The *Planck time*, which must be roughly the time it takes for something truly fundamental to happen, is 10^{-43} of a second. That is, the quickest thing we can experience still takes more than 10^{40} fundamental moments. A blink of an eye has more fundamental moments than there are atoms in Mount Everest. Even the fastest collision ever observed between two elementary particles fills more elementary moments than there are neurons•in the brains of all the people now alive. It is hard to avoid the conclusion that everything we observe may still be incredibly complicated on the fundamental Planck scale.

We can go on like this. There is a fundamental *Planck temperature*, which is likely to be the hottest anything can get. Compared with it, everything in our experience, even the interiors of stars, is barely above absolute zero. This means that, in terms of fundamental things the universe we observe is frozen. We begin to get the feeling that we know as much about nature and its potential phenomena as a penguin knows of the effects of forest fire, or of nuclear fusion. This is not just an analogy – it is our real situation. We know that all materials melt when raised to a high enough temperature. If a region of the world were raised to the Planck temperature, the very structure of the geometry of space would melt. The only hope we have of experiencing such an event is by peering into our past, for what is usually called the big bang is, in fundamental terms, the big freeze. What caused our world to exist was probably not so much an explosion as an event that caused a region of the universe to cool drastically and freeze. To understand space and time in their natural terms, we have to imagine what was there before everything around us froze.

So, our world is incredibly big, slow and cold compared with the fundamental world. Our job is to remove the prejudices and blinkers imposed by our parochial perspective and imagine space and time in their own terms, on their natural scale. We do have a very powerful toolkit that enables us to do this, consisting of the theories we have so far developed. We must take the theories that we trust the most, and tune them as best we can to give us a picture of the Planck scale. The story I am telling in this book is based on what we have learned by doing this.

In the earlier chapters I argued that our world cannot be understood as a collection of independent entities living in a fixed, static background of space and time. Instead, it is a network of relationships the properties of every part of which are determined by its relationships to the other parts. In this chapter we have learned that the relations that make up the world are causal relations. This means that the world is not made of stuff, but of processes by which things happen. Elementary particles are not static objects just sitting there, but processes carrying little bits of information between events at which they interact, giving rise to new processes. They are much more like the elementary operations in a computer than the traditional picture of an eternal atom.

We are very used to imagining that we see a three-dimensional world when we look around ourselves. But is this really true? If we keep in mind that what we see is the result of photons impinging on our eyes, it is possible to imagine our view of the world in a quite different way. Look around and imagine that you see each object as a consequence of photons having just travelled from it to you. Each object you see is the result of a process by which information travelled to you in the shape of a collection of photons. The farther away the object is, the longer it took the photons to travel to you. So when you look around you do not see space – instead, you are looking back through the history of the universe. What you are seeing is a slice through the history of the world. Everything you see is a bit of information brought to you by a process which is a small part of that history.

The whole history of the world is then nothing but the story of huge numbers of these processes, whose relationships are continually evolving. We cannot understand the world we see around us as something static. We must see it as something created, and under continual recreation, by an enormous number of processes acting together. The world we see around us is the collective result of all those processes. I hope this doesn't seem too mystical. If I have written this book well then, by the end of it, you may see that the analogy between the history of the universe and the flow of information in a computer is the most rational, scientific analogy I could make. What is mystical is the picture of the world as existing in an eternal three-dimensional space, extending in all directions as far as the mind can imagine. The idea of space going on and on for ever has nothing to do with what we see. When we look out, we are looking back in time through the history of the universe, and after not too long we come to the big bang. Before that there may be nothing to see – or, at the very least, if there is something it will most likely look nothing like a world suspended in a static three-dimensional space. When we imagine we are seeing into an infinite three-dimensional space, we are falling for a fallacy in which we substitute what we actually see for an intellectual construct. This is not only a mystical vision, it is wrong.

II
WHAT WE HAVE
LEARNED

CHAPTER 5

BLACK HOLES AND HIDDEN REGIONS

In the cultural iconography of our time, black holes have become mythic objects. In science fiction novels and films they often evoke images of death and transcendence, recalling the irreversibility of certain passages and the promise of our eventual emergence into a new universe. I am not a very good actor, but I was once asked by a friend, the director Madeline Schwartzman, to act in one of her films. Luckily I got to play a physics professor giving a lecture on black holes. In the film, called *Soma Sema*, the myth of Orpheus is merged with two major scientific and technological themes of our time: total nuclear war and black holes. Orpheus, my student, seeks through her music to be an exception to all three versions of the irreversible.

Among those of us who think about space and time professionally, black holes play a central role. A whole subculture of astronomers is devoted to understanding how they form and how to find them. By now, dozens of candidate black holes have been observed. But what is most exciting is that there are probably vast numbers of them out there. Many if not most galaxies, including our own, seem to have an enormous black hole at their centre, with a mass millions of times that of our Sun. And there is evidence, both observational and theoretical, that a small fraction of stars end their lives as black holes. A typical galaxy such as ours could well contain tens or even hundreds of millions of these stellar black holes. So black holes are out there, and interstellar travellers of the far future will have to be careful to avoid them. But beyond the fascination they hold for astronomers, black holes are important to science for other reasons. They are a central object of study for those of us who work on quantum gravity. In a sense, black holes are microscopes of infinite power which make it possible for us to see the physics that operates on the Planck scale.

Because they feature prominently in popular culture, almost everyone knows roughly what a black hole is. It is a place where gravity is so strong that the velocity required to escape from it is greater than the speed of light. So no light can emerge from it, and neither can anything else. We can understand this in terms of the notion of causal structure we introduced in the last chapter. A black hole contains a great concentration of mass that causes the light cones to tip over so far that the light moving away from the black hole actually gets no farther from it (Figure 13). So the surface of a black hole is like a one-way mirror: light moving towards it can pass into it, but no light can escape from it. For this reason the surface of a black hole is called the *horizon*. It is the limit of what observers outside the black hole can see.

I should emphasize that the horizon is not the surface of the object that formed the black hole. Rather it is the boundary of the region that is capable of sending light out into the universe. Light emitted by any body inside the horizon is trapped and cannot get any farther than the horizon. The object that formed the black hole is rapidly compressed, and according to general relativity it quickly reaches infinite density.

Behind the horizon of a black hole is a part of the universe made up of causal processes that go on, in spite of the fact that we receive no information from them. Such a region is called a *hidden region*. There are at least a billion billion black holes in the universe, so there are quite a lot of hidden regions that are invisible to us, or to any other observer. Whether a region is hidden or not depends in part on the observer. An observer who falls into a black hole will see things that her friends who stay outside will never see. In Chapter 2 we found that different observers may see different parts of the universe in

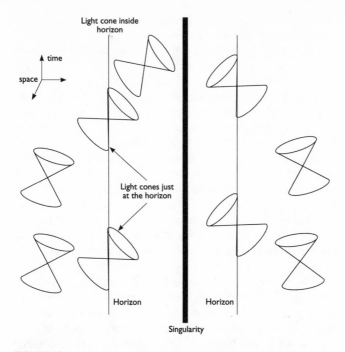

FIGURE 13

Light cones in the vicinity of a black hole. The solid black line is the singularity where the gravitational field is infinitely strong. The dotted lines are the horizons, consisting of light rays that stay the same distance from the singularity. Light cones just at the horizon are tilted to show that a light ray trying to move away from the black hole just stays at the same distance and travels along the horizon. A light cone inside the horizon is tilted so far that any motion into the future brings one closer to the singularity.

their past. The existence of black holes means that this is not just a question of waiting long enough for light from a distant region to reach us. We could be right next to a black hole, yet never be able to see what observers inside it can see, however long we waited.

All observers have their own hidden region. The hidden region of each observer consists of all those events that they will not be able to receive information from, no matter how long they wait. Each hidden region will include the interiors of all the black holes in the universe, but there may be other regions hidden as well. For example, if the rate at which the universe expands increases with time, there will be regions of the universe from which we shall never receive light signals, no matter how long we wait. A photon from such a region may be travelling in our direction at the speed of light, but because of the increase in the rate of the expansion of the universe it will always have more distance to travel towards us than it has travelled so far. As long as the expansion continues to accelerate, the photon will never reach us. Unlike black holes, the hidden regions produced by the acceleration of the expansion of the universe depend on the history of each observer. For each observer there is a hidden region, but they are different for different observers.

This raises an interesting philosophical point, because objectivity is usually assumed to be connected with observer independence. It is commonly assumed that anything that is observer dependent is subjective, meaning that it is not quite real. But the belief that observer dependence rules out objectivity is a residue of an older philosophy, usually associated with the name of Plato, according to which truth resides not in our world but in an imaginary world consisting of all ideas which are eternally true. According to this philosophy, anybody could have access to any truth about the world, because the process of finding truth was held to be akin to a process of remembering, rather than observing. This philosophy is hard to square with Einstein's general theory of relativity because, in a universe defined by that theory, something may be both objectively true and at the same time knowable only by some observers and not others. So 'objectivity' is not the same as 'knowable by all'. A weaker, less stringent interpretation is required: that all those observers who are in a position to ascertain the truth or falsity of some observation should agree with one another.

The hidden region of any observer has a boundary that divides the part of the universe they can see from the part they cannot. As with a black hole, this boundary is called the horizon. Like the invisible regions, horizons are observer dependent concepts. For any observer who remains outside it, a black hole has a horizon – the surface that divides the region from which light cannot escape from the rest of the universe. Light leaving a point just inside the horizon of the black hole will be pulled inexorably into the interior; light just outside the horizon will be able to escape (Figures 13 and 14). Although the horizon of a black hole is an observer dependent concept, there are a large number of observers who share that horizon: all those who are outside that black hole. So the horizon of a black hole is an objective property. But it is not a horizon for all observers, because any observer who falls through it will be able to see inside. And an observer who crosses the horizon of a black hole will become invisible to observers who remain outside.

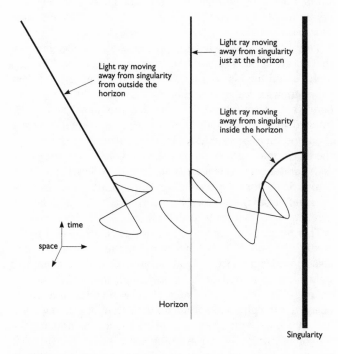

FIGURE 14
The paths of three light rays that move away from the singularity. They start just inside, outside and just at the horizon.

It helps to know that horizons are themselves surfaces of light. They are made up of those light rays that just fail to reach the observer (Figure 14). The horizon of a black hole is a surface of light that has begun to move outwards from the black hole but, because of the black hole's gravitational field, fails to get any farther from its centre. Think of the horizon as a curtain made of photons. Photons leaving from any point just inside the horizon are drawn inwards, even if they were initially moving away from the centre of the black hole.

On the other hand, a photon that starts just outside the horizon of a black hole will reach us, but it will be delayed because light cones near the horizon are tilted almost so far that no light can escape. The closer to the horizon the photon starts, the longer will be the delay. The horizon is the point where the delay becomes infinite – a photon released there never reaches us.

This has the following interesting consequence. Imagine that we are floating some distance from a black hole. We drop a clock into the black hole, which sends us a pulse of light every thousandth of a second. We receive the signal and convert it to sound. At first we hear the signal as a high-pitched tone, as we receive the signals at a frequency of a thousand times a second. But as the clock nears the horizon of the black hole, each signal is delayed more and more by the fact that it takes a little more time for each successive pulse to arrive. So the tone we hear falls in pitch as the clock nears the horizon. Just as the clock crosses the horizon, the pitch falls to zero, and after that we hear nothing.

This means that the frequency of light is decreased by its having to climb out from the region near the horizon. This can also be understood from quantum theory, as the frequency of light is proportional to its energy, and, just as it takes us energy to climb a flight of stairs, it takes a certain amount of energy for the photon to climb up to us from its starting point just outside the black hole. The closer to the horizon the photon begins its flight, the more energy it must give up as it travels to us. So the closer to the horizon it starts, the more its frequency will have decreased by the time it reaches us. Another consequence is that the wavelength of the light is lengthened as the frequency is decreased. This is because the wavelength of light is always inversely proportional to its frequency. As a result, if the frequency is diminished, the wavelength must be increased by the same factor.

But this means that the black hole is acting as a kind of microscope. It is not an ordinary microscope, as it does not act by enlarging images of objects. Rather, it acts by stretching wavelengths of light. But nevertheless, this is very useful to us. For suppose that at very short distances space has a different nature than the space we see looking around at ordinary scales. Space would then look very different from the simple three-dimensional Euclidean geometry that seems to suffice to describe the immediately perceptible world. There are various possibilities, and we shall be discussing these in later chapters. Space may be discrete, which means that geometry comes in bits of a certain absolute size. Or there may be quantum uncertainty in the very geometry of space. Just as electrons cannot be localized at precise points in the atom, but are forever dancing around the nucleus, the geometry of space may itself be dancing and fluctuating.

Ordinarily we cannot see what is happening on very small length scales. The reason is that we cannot use light to look at something which is smaller than the wavelength of that light. If we use ordinary light, even the best microscope will not resolve any object smaller than a few thousand times the diameter of an atom, which is the wavelength of the visible part of the spectrum of light. To see smaller objects we can use ultraviolet light, but no microscope in existence, not even one that uses electrons or protons in place of light, can come anywhere near the resolution required to see the quantum structure of space.

But black holes offer us a way around this problem. Whatever is happening on very small scales near the horizon of the black hole will be enlarged by the effect whereby the wavelengths of light are stretched as the light climbs up to us. This means that if we can observe light coming from very close to the horizon of a black hole, we may be able to see the quantum structure of space itself.

Unfortunately, it has so far proved impractical to make a

black hole, so no one has been able to do this experiment. But since the early 1970s several remarkable predictions have been made about what we would see if we could detect light coming from the region just outside a black hole. These predictions constitute the first set of lessons to have come from combining relativity and quantum mechanics. The next three chapters are devoted to them.

....

CHAPTER 6

ACCELERATION AND HEAT

To really understand what a black hole is like, we must imagine ourselves looking at one up close. What would we see if we were to hover just outside the horizon of a black hole (Figure 15)? A black hole has a gravitational field, like a planet or a star. So to hover just above its surface we must keep our rocket engines on. If we turn off our engines we shall go into a free fall that will quickly take us through the horizon and into the interior of the black hole. To avoid this we must continually accelerate to keep ourselves from being pulled down by the black hole's gravitational field. Our situation is similar to that of an astronaut in a lunar lander hovering over the surface of the Moon; the main difference is that we do not see a surface below us. Anything that falls towards the black hole accelerates past us as it falls towards the horizon, just below us. But we do not see the horizon because it is made up of photons that cannot reach us, even though they are moving in our direction. They are held in place by the black hole's gravitational field. So we see light coming from things between us and the horizon, but we see no light from the horizon itself.

You may well think there is something wrong with this. Are we really able to hover over a surface made of photons which never reach us, even though they are moving in our direction? Surely this contradicts relativity, which says that nothing can outrun light? This is true, but there is some fine print. If you are an inertial observer (that is, if you are moving at constant

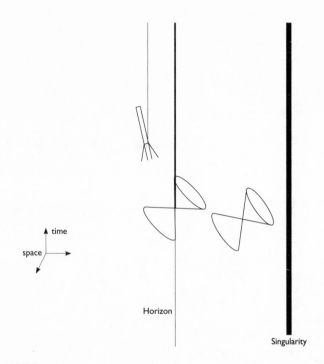

FIGURE 15

A rocket hovering just outside the horizon of a black hole. By keeping its engines on, the rocket can hover a fixed distance over the horizon.

speed, without accelerating) light will always catch up with you. But if you continually accelerate, then light, if it starts out from a point sufficiently far behind you, will never be able to catch you up. In fact this has nothing to do with a black hole. Any observer who continually accelerates, anywhere in the universe, will find themself in a situation rather like that of someone hovering just above the horizon of a black hole. We can see this from Figure 16: given enough of a head start, an accelerating observer can outrun photons. So an accelerating observer has a hidden region simply by virtue of the fact that photons cannot catch up with her. And she has a horizon, which is the boundary of her hidden region. The boundary separates those photons that will catch up with her from those that will not. It is made up of photons which, in spite of their moving at the speed of light, never come any closer to her. Of course, this horizon is due entirely to the acceleration. As soon as the observer turns off her engines and moves inertially, the light from the horizon and beyond will catch her up.

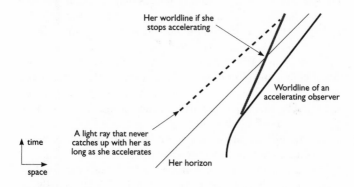

FIGURE 16
We see in bold the worldline of an observer who is constantly accelerating. She approaches but never passes the path of a light ray, which is her horizon since she can see nothing beyond it provided she continues to accelerate. Behind the horizon we see the path of a light ray that never catches up with her. We also see what her trajectory will be if she stops accelerating: she will then pass through her horizon and be able to see what lies on the other side.

This may seem confusing. How can an observer continually accelerate if it is not possible to travel faster than light? Rest assured that what I am saying in no way contradicts relativity. The reason is that while the continually accelerating observer never goes faster than light, she approaches ever closer to that limit. In each interval of time the same acceleration results in smaller and smaller increases in velocity. She comes ever closer to the speed of light, but never reaches it. This is because her mass increases as she approaches the speed of light. Were her speed to match that of light, her mass would become infinite. But one cannot accelerate an object that has infinite mass, hence one cannot accelerate an object to the speed of light or beyond. At the same time, relative to our clocks, her time seems to run slower and slower as her speed approaches, but never reaches, that of light. This goes on for as long as she keeps her engines on and continues to accelerate.

What we are describing here is a metaphor which is very useful for thinking about black holes. An observer hovering just above the surface of a black hole is in many ways just like an observer who is continually accelerating in a region far from any star or black hole. In both cases there is an invisible region whose boundary is a horizon. The horizon is made of light that travels in the same direction as the observer, but never comes any closer to her. To fall through the horizon, the observer has only to turn off her engines. When she does, the light that forms the horizon catches her up and she passes into the hidden region behind it.

But while the situation of an accelerating observer is analogous to that of an observer just outside a black hole, in some ways her situation is simpler. So in this chapter we shall take a small detour and consider the world as seen by an observer who constantly accelerates. This will teach us the concepts we need to understand the quantum properties of a black hole.

Of course, the two situations are not completely analogous. They differ in that the black hole's horizon is an objective property of the black hole, which is seen by many other observers. However, the invisible region and horizon of an accelerating observer are consequences only of her acceleration, and are seen only by her. Still, the metaphor is very useful. To see why, let us ask a simple question: what does our continually accelerating observer see when she looks around her?

Assume that the region she accelerates through is completely empty. There is no matter or radiation anywhere nearby – there is nothing but the vacuum of empty space. Let us equip our accelerating observer with a suite of scientific instruments, like the ones carried by space probes: particle detectors, thermometers, and so on. Before she turns on her engines she sees nothing, for she is in a region where space is truly empty. Surely turning on her engines does not change this?

In fact it does. First she will experience the normal effect of acceleration, which is to make her feel heavy, just as though she were all of a sudden in a gravitational field. The equivalence between the effects of acceleration and gravity is familiar from the experiences of life and from the science fiction fantasies of artificial gravity in rotating space stations. It is also the most basic principle of Einstein's general theory of relativity. Einstein called this the *equivalence principle*. It states that if one is in a windowless room, and has no contact with the outside, it is impossible to tell if one's room is sitting on the surface of the Earth, or is far away in empty space but accelerating at a rate equal to that by which we see objects fall towards the Earth.

But one of the most remarkable advances of modern theoretical physics has been the discovery that acceleration has another effect which seems at first to have nothing at all to do with gravity. This new effect is very simple: as soon as she accelerates, our observer's particle detectors will begin to register, in spite of the fact that, according to a normal observer who is not accelerating, the space through which she is travelling is empty. In other words, she will not agree with her non-accelerating friends on the very simple question of whether the space through which they are travelling is empty. The observers who do not accelerate see a completely empty space – a vacuum. Our accelerating observer sees herself as travelling through a region filled with particles. These effects have nothing to do with her engines – they would still be appar-ent if she was being accelerated by being pulled by a rope. They are a universal consequence of her acceleration through space.

Even more remarkable is what she will see if she looks at her thermometer. Before she began accelerating it read zero, because temperature is a measure of the energy in random motion, and in empty space there is nothing to give a non-zero temperature. Now the thermometer registers a temperature, even though all that has changed is her acceleration. If she experiments, she will find that the temperature is propor-tional to her acceleration. Indeed, all her instruments will behave exactly as if she were all of a sudden surrounded by a gas of photons and other particles, all at a temperature which increases in proportion to her acceleration.

I must stress that what I am describing has never been observed. It is a prediction that was first made in the early 1970s by a brilliant young Canadian physicist, Bill Unruh, who was then barely out of graduate school. What he found was that, as a result of quantum theory and relativity, there must be a new effect, never observed but still universal, whereby anything which is accelerated must experience itself to be embedded in a hot gas of photons, the temperature of which is proportional to the acceleration. The exact relation between temperature T and acceleration a is known, and is given by a famous formula first derived by Unruh. This formula is so simple we can quote it here:

$$T = a(\bar{h}/2\pi c)$$

The factor $\bar{h}/2\pi c$, where $\bar{h}$ is Planck's constant and c is the speed of light, is small in ordinary units, which means that the effect has so far escaped experimental confirmation. But it is not inaccessible, and there are proposals to measure it by accelerating electrons with huge lasers. In a world without quantum theory, Planck's constant would be zero and there would be no effect. The effect also goes away when the speed of light goes to infinity, so it would also vanish in Newtonian physics.

This effect implies that there is a kind of addendum to Einstein's famous equivalence principle. According to Einstein, a constantly accelerating observer should be in a situation just like an observer sitting on the surface of a planet. Unruh told us that this is true only if the planet has been heated to a temperature that is proportional to the acceleration.

What is the origin of the heat detected by an accelerating observer? Heat is energy, which we know cannot be created nor destroyed. Thus if the observer's thermometer heats up there must be a source of the energy. So where does it come from? The energy comes from the observer's own rocket engines. This makes sense, for the effect is present only as long as the observer is accelerating, and this requires a constant input of energy. Heat is not only energy, it is energy in random motion. So we must ask how the radiation measured by an accelerating particle detector becomes randomized. To understand this we have to delve into the mysteries of the quantum theoretic description of empty space.

According to quantum theory, no particle can sit exactly still for this would violate Heisenberg's uncertainty principle. A particle that remains at rest has a precise position, for it never moves. But for the same reason it has also a precise momentum, namely zero. This also violates the uncertainty principle: we cannot know both position and momentum to arbitrary precision. The principle tells us that if we know the position of a particle with absolute precision we must be completely ignorant of the value of its momentum, and vice versa. As a consequence, even if we could remove all the energy from a particle, there would remain some intrinsic random motion. This motion is called the *zero point motion*.

What is less well known is that this principle also applies to the fields that permeate space, such as the electric and magnetic fields that carry the forces originating in magnets and electric currents. In this case the roles of position and momentum are played by the electric and magnetic fields. If one measures the precise value of the electric field in some region, one must be completely ignorant of the magnetic field, and so on. This means that if we measure both the electric and magnetic fields in a region we cannot find that both are zero. Thus, even if we could cool a region of space down to zero temperature, so that it contained no energy, there would still be randomly fluctuating electric and magnetic fields. These are called the *quantum fluctuations* of the vacuum. These quantum fluctuations cannot be detected by any ordinary instrument, sitting at rest, because they carry no energy, and only energy can register its presence in a detector. But the amazing thing is that they can be detected by an accelerating detector, because the acceleration of the detector provides a source of energy. It is exactly these random quantum fluctuations that raise the temperature of the thermometers carried by our accelerating observer.

This still does not completely explain where the randomness comes from. It turns out to have to do with another central concept in quantum theory, which is that there are non-local correlations between quantum systems. These correlations can be observed in certain special situations such as the Einstein–Podolsky–Rosen experiment. In this experiment two photons are created together, but travel apart at the speed of light. But when they are measured it is found that their properties are correlated in such a way that a complete description of either one of them involves the other. This is true no matter how far apart they travel (Figure 17). The photons that make up the vacuum electric and magnetic fields come in pairs that are correlated in exactly this way. What is more, each photon detected by our accelerating observer's thermometer is correlated with one that is beyond her horizon. This means that part of the information she would need if she wanted to give a complete description of each photon she sees is inaccessible to her, because it resides in a photon that is in her hidden region. As a result, what she observes is intrinsically random. As with the atoms in a gas, there is no way for her to predict exactly how the photons she observes are moving. The result is that the motion she sees is random. But random motion is, by definition, heat. So the photons she sees are hot!

Let us follow this story a bit further. Physicists have a measure of how much randomness is present in any hot system. It is called entropy, and is a measure of exactly how much disorder or randomness there is in the motion of the atoms in any hot system. This measure can be applied also to photons. For example, we can say that the photons coming from the test pattern on my television, being random, have more entropy than the photons that convey *The X Files* to my eyes. The photons detected by the accelerating detector are random, and so do have a finite amount of entropy.

Entropy is closely related to the concept of information. Physicists and engineers have a measure of how much information is available in any signal or pattern. The information carried by a signal is defined to be equal to the number of yes/no questions whose answers could be coded in that

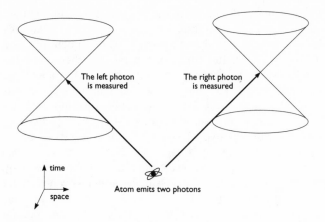

FIGURE 17

The Einstein–Podolsky–Rosen (EPR) experiment. Two photons are created by the decay of an atom. They travel in opposite directions, and are then measured at two events which are outside each other's light cones. This means that no information can flow to the left event about which measurement the right observer chooses to make. Nevertheless, there are correlations between what the left observer sees and what the right observer chooses to measure. These correlations do not transmit information faster than light because they can be detected only when the statistics from the measurements on each side are compared.

signal. In our digital world, most signals are transmitted as a sequence of bits. These are sequences of ones and zeroes, which may also be thought of as sequences of yeses and noes. The information content of a signal is thus equal to the number of bits, as each bit may be coding the answer to a yes/no question. A megabyte is then precisely a measure of information in this sense, and a computer with a memory of, say, 100 megabytes can store 100 million bytes of information. As each byte contains 8 bits, and each corresponds to the answer to a single yes/no question, this means that the 100 megabyte memory can store the answers to 800 million yes/no questions.

In a random system such as a gas at some non-zero temperature, a large amount of information is coded in the random motion of the molecules. This is information about the positions and motions of the molecules that does not get specified when one describes the gas in terms of quantities such as density and temperature. These quantities are averaged over all the atoms in the gas, so when one talks about a gas in this way most of the information about the actual positions and motions of the molecules is thrown away. The entropy of a gas is a measure of this information – it is equal to the number of yes/no questions that would have to be answered to give a precise quantum theoretic description of all the atoms in the gas.

Information about the exact states of the hot photons seen by the accelerating observer is missing because it is coded in the states of the photons in her hidden region. Because the randomness is a result of the presence of the hidden region, the entropy should incorporate some measure of how much of the world cannot be seen by the accelerating observer. It should have something to do with the size of her hidden region. This is almost right; it is actually a measure of the size of the boundary that separates her from her hidden region. The entropy of the hot radiation she observes as a result of her acceleration turns out to be exactly proportional to the area of her horizon! This relationship between the area of a horizon and entropy was discovered by a Ph.D. student named Jacob Bekenstein, who was working at Princeton at about the time that Bill Unruh made his great discovery. Both were students of John Wheeler, who a few years before had given the black hole its name. Bekenstein and Unruh were in a long line of remarkable students Wheeler trained, which included Richard Feynman.

What those two young physicists did remains the most important step yet made in the search for quantum gravity. They gave us two general and simple laws, which were the first physical predictions to come from the study of quantum gravity. They are:

- *Unruh's law* Accelerating observers see themselves as embedded in a gas of hot photons at a temperature proportional to their acceleration.

- *Bekenstein's law* With every horizon that forms a boundary separating an observer from a region which is hidden from them, there is associated an entropy which measures the amount of information which is hidden behind it. This entropy is always proportional to the area of the horizon.

These two laws are the basis for our understanding of quantum black holes, as we shall see in the next chapter.

....

CHAPTER 7

BLACK HOLES ARE HOT

The reason why we have been considering an accelerating observer is that her situation is very similar to that of an observer hovering just above the horizon of a black hole. So the two laws we found at the end of the last chapter, Unruh's law and Bekenstein's law, can be applied to tell us what we see as we hover over a black hole. Applying the analogy, we can predict that an observer outside the black hole will see themself as embedded in a gas of hot photons. Their temperature must be related to the acceleration the engines need to deliver to keep the spacecraft hovering a fixed distance above the horizon. Furthermore, the photons that this observer detects will be randomized because a complete description of them will require information that is beyond the horizon, coded in correlations between the photons she sees and photons that remain beyond the horizon (Figure 18). To measure this missing information she will attribute an entropy to the black hole. And this entropy will turn out to be proportional to the area of the horizon of the black hole.

Although the analogy is very useful, there is an important difference between the two situations. The temperature and entropy measured by the accelerating observer are consequences of her motion alone. If she turns off her engines, the photons making up her horizon will catch up with her. She can then see into her hidden region. She no longer sees a hot gas of photons, so she measures no temperature. There is no missing information as she sees only empty space, which is

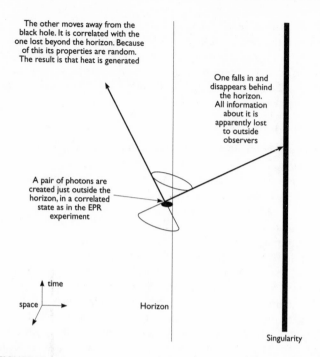

The other moves away from the black hole. It is correlated with the one lost beyond the horizon. Because of this its properties are random. The result is that heat is generated

One falls in and disappears behind the horizon. All information about it is apparently lost to outside observers

A pair of photons are created just outside the horizon, in a correlated state as in the EPR experiment

time

space

Horizon

Singularity

FIGURE 18

Radiation from black holes, as discovered by Stephen Hawking. The photon that travels away from the black hole has random properties and motion because it is correlated, as in one of the photons in Figure 17, with the one lost behind the horizon. Because observers outside the horizon cannot recover the information that the infalling photon carries, the outmoving photon appears to have a thermal motion, like a molecule in a hot gas. The result is that the radiation leaving the black hole has a non-zero temperature. It also has an entropy, which is a measure of the missing information.

consistent with the fact that there is no hidden region, so no horizon. But with the black hole there are an infinite number of observers who agree that there is a horizon, beyond which they cannot see. And this is not just a consequence of their motion, for all observers who do not fall through the horizon will agree that the black hole and its horizon are there. This means that all observers who are far from a black hole will agree that it has a temperature and an entropy.

For simple black holes, which do not rotate and have no electric charge, the values of the temperature and entropy can be expressed very simply. The area of the horizon of a simple black hole is proportional to the square of its mass, in Planck units. The entropy S is proportional to this quantity. In terms of Planck units, we have the simple formula

$$S = \frac{1}{4} A / \hbar G$$

where A is the area of the horizon, and G is the gravitational constant.

There is a very simple way to interpret this equation which is due to Gerard 't Hooft, who did important work in elementary particle physics – for which he won the 1999 Nobel Prize for Physics – before turning his attention to the problem of quantum gravity. He suggests that the horizon of a black hole is like a computer screen, with one pixel for every four Planck areas. Each pixel can be on or off, which means that it codes one bit of information. The total number of bits of information contained within a black hole is then equal to the total number of such pixels that it would take to cover the horizon. The Planck units are very, very tiny. It would take 10^{66} Planck area pixels to cover a single square centimetre. So an astrophysical black hole whose horizon has a diameter of several kilometres can contain a stupendous amount of information.

Entropy has another meaning besides being a measure of information. If a system has entropy, it will act in ways that are irreversible in time. This is because of the second law of thermodynamics, which says that entropy can only be created, not destroyed. If you shatter a teapot by dropping it on the floor, you have greatly increased its entropy – it will be very difficult to put it back together. In thermodynamics the irreversibility of a process is measured by an increase of entropy, because that measures the amount of information lost to random motion. But such information, once lost, can never be recovered, so the entropy cannot normally decrease. This is one way of expressing the second law of thermodynamics.

Black holes also behave in a way that is not reversible in

time, because things can fall into a black hole but nothing can come out of one. This turns out to have a consequence, first discovered by Stephen Hawking, for the area of the horizon of a black hole. He showed by a very elegant proof that the area of the horizon of a black hole can never decrease in time. So it was natural to suggest that the area of the horizon of a black hole is analogous to entropy, in that it is a quantity that can only increase in time. The great insight of Bekenstein was that this was not just an analogy. He argued that a black hole has real entropy, which he conjectured is proportional to the area of its horizon and measures the amount of information trapped beyond that horizon.

You may wonder why fifteen thousand other physicists were not able to take this step if it was based on a simple analogy, apparent to anyone who looked at the problem. The reason is that the analogy is not quite complete. For if nothing can come out of a black hole, then it has zero temperature. This is because temperature measures the energy in random motion, and if there is nothing in a box, there can be no motion of any kind, random or otherwise. But ordinarily a system cannot have entropy without it being hot. This is because the missing information results in random motion, which means there is heat. So, if black holes have entropy, there is a violation of the laws of thermodynamics.

So this seemed to be not a brilliant move, but the kind of misuse of analogy that characterizes the thinking of novices in any field. But a few people did take Bekenstein seriously, including Stephen Hawking, Paul Davies and Bill Unruh. The mystery was solved first by Hawking, who realized that if black holes were hot there was no contradiction with the laws of thermodynamics. By following a chain of reasoning roughly like the story above, he was able to show that an observer outside a black hole would see it to be at a finite temperature. Expressed in Planck units, the temperature T of a black hole is inversely proportional to its mass, m. This is a third law, Hawking's law:

$$T = k/m$$

The constant k is very small in normal units. As a result, astrophysical black holes have temperatures of a very small fraction of a degree. They are therefore much colder than the 2.7 degree microwave background. But a black hole of much smaller mass would be correspondingly hotter, even if it were smaller in size. A black hole the mass of Mount Everest would be no larger than a single atomic nucleus, but it would glow with a temperature greater than the centre of a star.

The radiation emitted by a black hole, called Hawking radiation, carries away energy. By Einstein's famous relationship between mass and energy, $E = mc^2$, this means that the radiation carries away mass as well. This implies that a black hole in empty space must lose mass, for there is no other source of energy to power the radiation it emits. The process by which a black hole radiates away its mass is called *black hole evaporation*. As a black hole evaporates, its mass decreases. But since its temperature is inversely proportional to its mass, as it loses mass it gets hotter. This will go on at least until the temperature becomes so hot that each photon emitted has roughly the Planck energy. At this point the mass of the black hole is itself roughly equal to the Planck mass, and its horizon is a few Planck lengths across. We have got down to the regime where quantum gravity holds sway. What happens to the black hole next could only be decided by a full quantum theory of gravity.

The evaporation of an astrophysical black hole is a very slow process. The evaporation rate, which depends on the temperature, is very low because the temperature itself is so low, initially. It would take a black hole the mass of the Sun about 10^{57} times the present age of the universe to evaporate. So this is not something we are going to observe soon. But the question of what happens at the end of black hole evaporation is one that fascinates those of us who think about quantum gravity. It is a subject in which it is easy to find paradoxes to mull over. For example, what happens to the information trapped inside a black hole? We have said that the amount of trapped information is proportional to the area of the horizon of the black hole. When the black hole evaporates, the area of its horizon decreases. Does this mean that the amount of trapped information decreases as well? If not, then there seems to be a contradiction, but if so we must explain how the information gets out, as it is coded in photons that are trapped behind the horizon.

Similarly, we can also ask whether the entropy of the black hole decreases as the area of the horizon shrinks. It seems that it must, as the two quantities are related. But surely this violates the second law of thermodynamics, which states that entropy can never decrease? One answer is that it need not, because the radiation emitted by the black hole has lots of entropy, which makes up for that lost by the black hole. The second law of thermodynamics requires only that the total entropy of the world never increase. If we include in this total the entropy of the black hole, then all the evidence we have is that the second law of thermodynamics still holds. When something falls into a black hole the outside world may lose some entropy, but the increase in the entropy of the black hole will more than make up for it. On the other hand, if the black hole radiates it loses surface area and hence entropy, but the entropy of the outside world will increase to make up for it.

The result of all this is at the same time very satisfying and deeply puzzling. It is satisfying because the study of black holes has led to a beautiful extension of the laws of thermodynamics. It seemed at first that black holes would violate the laws of thermodynamics. But eventually we realized that if black holes themselves have entropy and temperature, then the laws of thermodynamics would remain true. What is puzzling is that in most circumstances entropy is a measure of missing information. In classical general relativity a black hole is not something complicated: it is described by a few numbers such as its mass and electric charge. But if it has entropy there must be some missing information. The classical theory of black holes gives us no clue as to what that information is about. Nor do the calculations by Bekenstein, Hawking and Unruh give us any hint about what it might be.

But if there is no clue from the classical theory as to the nature of the missing information, there is only one possibility, which is that we need the quantum theory of a black hole to reveal it to us. If we could understand a black hole as a purely quantum system, then its entropy would have to turn out to include some information about itself that is evident only at the quantum level. So we may now pose a question which could be answered only if we have a quantum theory of gravity. What is the nature of the information trapped in a quantum black hole? Keep this in mind as we go ahead and explore the different approaches to quantum gravity, for a good test of a theory of quantum gravity is how well it is able to answer this question.

....

CHAPTER 8

AREA AND INFORMATION

At the beginning of the twentieth century, few physicists believed in atoms. Now there are few educated people who do not believe in them. But what about space? If we take a bit of space, say a cube 1 centimetre on each side, we can divide each side in two to give eight smaller pieces of space. We can divide each of these again, and so on. With matter there is a limit to how small we can divide something, for at some point we are left with individual atoms. Is the same true of space? If we continue dividing, do we eventually come to a smallest unit of space, some smallest possible volume? Or can we go on for ever, dividing space into smaller and smaller bits, without ever having to stop? All three of the roads I described in the Prologue favour the same answer to this question: that there is indeed a smallest unit of space. It is much smaller than an atom of matter, but nevertheless, as I shall describe in this chapter and the next three, there are good reasons to believe that the continuous appearance of space is as much an illusion as the smooth appearance of matter. When we look on a small enough scale, we see that space is made of things that we can count.

Perhaps it is hard to visualize space as something discrete. After all, why can something not be made to fit into half the volume of the smallest unit of space? The answer is that this is the wrong way to think, for to pose this question is to presume that space has some absolute existence into which things can fit. To understand what we mean when we say that space is discrete, we must put our minds completely into the relational way of thinking, and really try to see and feel the world around us as nothing but a network of evolving relationships. These relationships are not among things situated in space – they are among the events that make up the history of the world. The relationships define the space, not the other way round.

From this relational point of view it makes sense to say that the world is discrete. Actually it is easier, because then we have to conceive of only a finite number of events. It is harder to visualize a smooth space constructed from a network of relationships, as this would require there to be an infinite number of relationships between the events in any volume of space, however small that volume. Even if we had no other evidence (and we do), the fact that it makes the relational picture of spacetime so much easier to think about would be reason enough to imagine both space and time as discrete.

Of course, so far no one has ever observed an atom of space. Nor have any of the predictions that follow from the theories that predict that space is discrete been tested experimentally. So how is it that many physicists have already come to believe that space is discrete? This is indeed a good question, to which there is a good answer: the present situation is in some ways analogous to the period during which most physicists became convinced of the existence of atoms, during the twenty years spanning the last decade of the nineteenth century and the first decade of the twentieth. The first experiments that can be said to have detected atoms, which used the first, primitive elementary particle accelerators, were not done until just after this period, in 1911/12. By then most physicists were already convinced of the existence of atoms.

Presently we are in a crucial period during which the laws of physics are being rewritten – just as they were between 1890 and 1910, when the revolutions in twentieth-century physics that led to relativity and quantum physics began. The crucial arguments that led people to accept the existence of atoms were formulated during that period to resolve the paradoxes and contradictions that followed from the assumption that matter and radiation were continuous. The experiments that detected atoms came later because their very conceptualization required ideas that were invented as part of the same process. Had the experiments been done twenty years earlier, the results may not even have been interpreted as evidence for the existence of atoms.

The crucial arguments that convinced people of the existence of atoms had to do with understanding the laws governing heat, temperature and entropy – the part of physics called thermodynamics. Among the laws of thermodynamics are the second law, which we have already discussed, which states that entropy never decreases, and the so-called zeroth law, which states that when the entropy of a system is as high as possible, it has a single uniform temperature. Between them comes the first law, which asserts that energy is never created or destroyed.

During most of the nineteenth century most physicists did not believe in atoms. It is true that the chemists had found that different substances combine in fixed ratios, which was suggestive of the existence of atoms. But the physicists were not very impressed. Until 1905 most of them thought either that matter was continuous, or that the question of whether there were atoms or not lay outside science, because even if they existed atoms would be forever unobservable. These scientists developed the laws of thermodynamics in a form that made no reference to atoms or their motions. They did not believe the basic definitions of temperature and entropy that I introduced in earlier chapters: that temperature is a measure of the energy of random motion, and that entropy is a measure of information. Instead, they understood temperature and entropy as essential properties of matter: matter was just a continuous fluid or substance, and temperature and entropy were among its basic properties.

Not only did the laws of thermodynamics make no reference to atoms, but the nineteenth-century founders of the theory even believed there was a reason why there could be no relation between atoms and thermodynamics. This is because the second law, by saying that entropy increases towards the future, introduces an asymmetry in time. According to this law the future is different from the past because the future is the direction in which the entropy of the universe increases. On the other hand, these people reasoned that if there were atoms they would have to obey Newton's laws. But these laws are reversible in time. Suppose you were to make a movie of a set of particles interacting according to Newton's laws, and then show the movie twice to a group of physicists, once as it was made, and once running it backwards. As long as there were only a few particles in the movie, there is no way for the physicists to determine which was the right way for time to go.

Things are very different for large, macroscopic bodies. In the world we live in, the future is very different from the past, which is exactly what is captured in the law stating that entropy increases into the future. Because this seemed to contradict the fact that in Newton's theory the future and the past are reversible, many physicists refused to believe that matter is made of atoms until the first few decades of the twentieth century, when conclusive experimental proof was obtained for their existence.

The ideas that temperature is a measure of energy in random motion and entropy is a measure of information underlie what is called the statistical formulation of thermodynamics. According to this view ordinary matter is made out of enormous numbers of atoms. This means that one has to reason statistically about the behaviour of ordinary matter. According to the founders of *statistical mechanics*, as the idea was called, one could explain the apparent paradox about the direction of time by deriving the laws of thermodynamics from Newton's laws. The paradox was resolved by understanding that the laws of thermodynamics are not absolute: they describe what is most likely to happen, but there will always be a small probability of the laws being violated.

In particular, the laws assert that most of the time a large collection of atoms will evolve in such a way as to reach a more random – meaning more disorganized – state. This is just because the randomness of the interactions tends to wash out any organization or order that is initially present. But this need not happen, it is just what is most likely to happen. A

system which is very carefully prepared, or which incorporates structures that preserve a memory of what has happened
to it – such as a complex molecule such as DNA – can be seen
to evolve from a less ordered to a more ordered state.

The argument here is rather subtle, and it took several
decades for most physicists to be convinced. The originator of
the idea that entropy had to do with information and
probability, Ludwig Boltzmann, committed suicide in 1906,
which was before most physicists had accepted his arguments. (Whether his depression had anything to do with the
failure of his colleagues to appreciate his reasoning, Boltzmann's suicide had at least one far-reaching consequence:
it convinced a young physics student named Ludwig Witt-
genstein to give up physics and go to England to study
engineering and philosophy.) In fact, the arguments that
finally convinced most physicists of the existence of atoms
had just been published the year before by the then patent
office clerk Albert Einstein ('Same Einstein', as my physics
teacher used to say.) This argument had to do with fact that
the statistical point of view allowed the laws of thermodynamics to be violated from time to time. What Boltzmann
had found was that the laws of thermodynamics would be
exactly true for systems that contained an infinite number of
atoms. Of course, the number of atoms in a given system, such
as the water in a glass, is very large, but it is not infinite.
Einstein realized that for systems containing a finite number
of atoms the laws of thermodynamics would be violated from
time to time. Since the number of atoms in the glass is large,
these effects are small, but they still may in some circumstances be observed. By making use of this fact Einstein was
able to discover manifestations of the motions of atoms that
could be observed. Some of these had to do with the fact that a
grain of pollen, observed in a microscope, will dance around
randomly because it is being jiggled by atoms colliding with
it. As each atom has a finite size, and carries a finite amount of
energy, the jiggles that result when they collide with the grain
of pollen can be seen, even if the atoms themselves are far too
small to be seen.

The success of these arguments persuaded Einstein and a

few others, such as his friend Paul Ehrenfest, to apply the same reasoning to light. According to the theory published by James Clerk Maxwell in 1865, light consisted of waves travelling through the electromagnetic field, each wave carrying a certain amount of energy. Einstein and Ehrenfest wondered whether they could use Boltzmann's ideas to describe the properties of light on the inside of an oven.

Light is produced when the atoms in the walls of the oven heat up and jiggle around. Could the light so produced be said to be hot? Could it have an entropy and a temperature? What they found was profoundly puzzling to them and to everyone else at the time. They found that horrible inconsistencies would arise unless the light were in a sense also to consist of atoms. Each atom of light, or *quantum* as they called it, had to carry a unit of energy related to the frequency of the light. This was the birth of quantum theory.

I shall tell no more of this story, for it is indeed a very twisted one. Some of the results that Einstein and Ehrenfest employed in their reasoning had been found earlier by Max Planck, who had studied the problem of hot radiation five years earlier. It was in this work that the famous *Planck's constant* first appeared. But Planck was one of those physicists who believed neither in atoms nor in Boltzmann's work, so his understanding of his own results was confused and, in part, contradictory. He even managed to invent a convoluted argument that assured him that photons did not exist. For this reason the birth of quantum physics is more properly attributed to Einstein and Ehrenfest.

The moral of this story is that it was an attempt to understand the laws of thermodynamics that prompted two crucial steps in our understanding of atomic physics. These were the arguments that convinced physicists of the existence of atoms, and the arguments by which the existence of the photon were first uncovered. It was no coincidence that both these steps were taken by the same young Einstein in the same year.

We can now turn back to quantum gravity, and in particular to quantum black holes. For what we have seen in the last few chapters is that black holes are systems which may be described by the laws of thermodynamics. They have a temperature and an entropy, and they obey an extension of the law of increase of entropy. This allows us to raise several questions. What does the temperature of a black hole actually measure? What does the entropy of a black hole really describe? And, most importantly, why is the entropy of a black hole proportional to the area of its horizon?

The search for the meaning of temperature and entropy of matter led to the discovery of atoms. The search for the meaning of the temperature and entropy of radiation led to the discovery of quanta. In just the same way, the search for the meaning of the temperature and entropy of a black hole is now leading to the discovery of the atomic structure of space and time.

Consider a black hole interacting with a gas of atoms and photons. The black hole can swallow an atom or a photon. When it does so, the entropy of the region outside the black hole decreases because the entropy is a measure of information about that region, and if there are fewer atoms or photons there is less to know about the gas. To compensate, the entropy of the black hole must increase, otherwise the law that entropy can never decrease would be violated. As the entropy of the black hole is proportional to the area of its horizon, the result must be that the horizon expands a little.

And indeed, this is what happens. The process can also go the other way: the horizon can shrink a little, which means that the entropy of the black hole will decrease. To compensate, the entropy outside the black hole must increase. To accomplish this, photons must be created just outside the black hole – photons that comprise the radiation that Hawking predicted should be emitted by a black hole. The photons are hot, so they can carry the entropy that must be created to compensate for the fact that the horizon shrinks.

What is happening is that, to preserve the law that entropy does not decrease, a balance is being struck between, on the one hand, the entropy of atoms and photons outside the black hole, and, on the other, the entropy of the black hole itself. But notice that two very different things are being balanced. The entropy outside the black hole we understand in terms of the idea that matter is made out of atoms; it has to do with missing information. The entropy of the black hole itself seems to have nothing to do with either atoms or with information. It is a measure of a quantity which has to do with the geometry of space and time: it is proportional to the area of the black hole's event horizon.

There is something incomplete about a law which asserts a balance or an exchange between two very dissimilar things. It is as though we had two kinds of currency, the first of which was exchangeable into a concrete entity such as gold, while the other had no worth in terms other than paper. Suppose we were allowed to freely mix the two kinds of money in our bank accounts. Such an economy would be based on a contradiction, and could not survive for long. (In fact, communist governments experimented with two kinds of currency, one convertible into other currencies and one not, and discovered that the system is unstable in the absence of all sorts of complicated and artificial restrictions on the use of the two kinds of money.) Similarly, a law of physics that allows information to be converted into geometry, and vice versa, but gives no account of why, should not survive for long. There must be something deeper and simpler at the root of the equivalence.

This raises two profound questions:

- Is there an atomic structure to the geometry of space and time, so that the entropy of the black hole could be understood in exactly the same way that the entropy of matter is understood: as a measure of information about the motion of the atoms?

- When we understand the atomic structure of geometry will it be obvious why the area of a horizon is proportional to the amount of information it hides?

These questions have motivated a great deal of research since the mid-1970s. In the next few chapters I shall explain why there is a growing consensus among physicists that the answer to both questions must be 'yes'.

Both loop quantum gravity and string theory assert that there is an atomic structure to space. In the next two chapters we shall see that loop quantum gravity in fact gives a rather detailed picture of that atomic structure. The picture of the atomic structure one gets from string theory is presently incomplete but, as we shall see in Chapter 11, it is still impossible in string theory to avoid the conclusion that there must be an atomic structure to space and time. In Chapter 13 we shall discover that both pictures of the atomic structure of space can be used to explain the entropy and temperature of black holes.

But even without these detailed pictures there is a very general argument, based simply on what we have learned in the last few chapters, that leads to the conclusion that space must have an atomic structure. This argument rests on the simple fact that horizons have entropy. In previous chapters we have seen that this is common to both the horizons of black holes and to the horizon experienced by an accelerated observer. In each case there is a hidden region in which information can be trapped, outside the reach of external observers. Since entropy is a measure of missing information, it is reasonable that in these cases there is an entropy associated with the horizon, which is the boundary of the hidden region. But what was most remarkable is that the amount of missing information measured by the entropy had a very simple form. It was simply equal to one-quarter of the area of the horizon, in Planck units.

The fact that the amount of missing information depends on the area of the boundary of the trapped region is a very important clue. It becomes even more significant if we put this dependence together with the fact that spacetime can be understood to be structured by processes which transmit information from the past to the future, as we saw in Chapter 4. If a surface can be seen as a kind of channel through which information flows from one region of space to another, then the area of the surface is a measure of its capacity to transmit information. This is very suggestive.

It is also strange that the amount of trapped information is proportional to the area of the boundary. It would seem more natural for the amount of information that can be trapped in a region to be proportional to its volume, not to the area of its boundary. No matter what is on the other side of the boundary, trapped in the hidden region, it can contain the answer to only a finite number of yes/no questions per unit area of the boundary. This seems to be saying that a black hole, whose horizon has a finite area, can hold only a finite amount of information.

If this is the right interpretation of the results I described in the last chapter, it suffices to tells us that the world must be discrete, since whether a given volume of space is behind a horizon or not depends on the motion of an observer. For any volume of space we may want to consider, we can find an observer who accelerates away from it in such a way that that region becomes part of that observer's hidden region. This tells us that in that volume there could be no more information than the limit we are discussing, which is a finite amount per unit area of the boundary. If this is right, then no region can contain more than a finite amount of information. If the world really were continuous, then every volume of space would contain an infinite amount of information. In a continuous world it takes an infinite amount of information to specify the position of even one electron. This is because the position is given by a real number, and most real numbers require an infinite number of digits to describe them. If we write out their decimal expansion, it will require an infinite number of decimal places to write down the number.

In practice, the greatest amount of information that may be stored behind a horizon is huge – 10^{66} bits of information per square centimetre. No actual experiment so far comes close to probing this limit. But if we want to describe nature on the Planck scale, we shall certainly run into this limitation, as it allows us to talk about only one bit of information for every four Planck areas. After all, if the limit were one bit of information per square centimetre rather than per square Planck area, it would be quite hard to see anything because our eyes would then be able respond to at most one photon at a time.

Many of the important principles in twentieth-century

physics are expressed as limitations on what we can know. Einstein's principle of relativity (which was an extension of a principle of Galileo's) says that we cannot do any experiment that would distinguish being at rest from moving at a constant velocity. Heisenberg's uncertainty principle tells us that we cannot know both the position and momentum of a particle to arbitrary accuracy. This new limitation tells us there is an absolute bound to the information available to us about what is contained on the other side of a horizon. It is known as *Bekenstein's bound*, as it was discussed in papers Jacob Bekenstein wrote in the 1970s shortly after he discovered the entropy of black holes.

It is curious that, despite everyone who has worked on quantum gravity having been aware of this result, few seem to have taken it seriously for the twenty years following the publication of Bekenstein's papers. Although the arguments he used were simple, Jacob Bekenstein was far ahead of his time. The idea that there is an absolute limit to information which requires each region of space to contain at most a certain finite amount of information was just too shocking for us to assimilate at the time. There is no way to reconcile this with the view that space is continuous, for that implies that each finite volume can contain an infinite amount of information. Before Bekenstein's bound could be taken seriously, people had to discover other, independent reasons why space should have a discrete, atomic structure. To do this we had to learn to do physics at the scale of the smallest possible things.

....

CHAPTER 9

HOW TO COUNT SPACE

The first approach to quantum gravity that yielded a detailed description of the atomic structure of space and spacetime was *loop quantum gravity*. The theory offers more than a picture: it makes precise predictions about what would be observed were it possible to probe the geometry of space at distances as short as the Planck scale.

According to loop quantum gravity, space is made of discrete atoms each of which carries a very tiny unit of volume. In contrast to ordinary geometry, a given region cannot have a volume which is arbitrarily big or small – instead, the volume must be one of a finite set of numbers. This is just what quantum theory does with other quantities: it restricts a quantity that is continuous according to Newtonian physics to a finite set of values. This is what happens to the energy of an electron in an atom, and to the value of the electric charge. As a result, we say that the volume of space is predicted to be quantized.

One consequence of this is that there is a smallest possible volume. This minimum volume is minuscule – about 10^{99} of them would fit into a thimble. If you tried to halve a region of this volume, the result would not be two regions each with half that volume. Instead, the process would create two new regions which together would have more volume than you started with. We describe this by saying that the attempt to measure a unit of volume smaller than the minimal size alters the geometry of the space in a way that allows more volume to be created

Volume is not the only quantity which is quantized in loop

quantum gravity. Any region of space is surrounded by a boundary which, being a surface, will have an area, and that area will be measured in square centimetres. In classical geometry a surface can have any area. In contrast, loop quantum gravity predicts that there is a smallest possible area. As with volume, the theory limits the possible areas a surface can have to a finite set of values. In both cases the jumps between possible values are very small, of the order of the square and cube of the Planck length. This is why we have the illusion that space is continuous.

These predictions could be confirmed or refuted by measurements of the geometry of things made on the Planck scale. The problem is that because the Planck scale is so small, it is not easy to make these measurements – but it is not impossible, as I shall describe in due course.

In this chapter and the next I shall tell the story of how loop quantum gravity developed from a few simple ideas into a detailed picture of space and time on the shortest possible scales. The style of these chapters will be rather more narrative than the others, as I can describe from personal experience some of the episodes in the development of the theory. I do this mainly to illustrate the complicated and unexpected ways in which a scientific idea can develop. This can only be communicated by telling stories, but I must emphasize that there are many stories. My guess is that the inventors of string theory have better stories, with more human drama. I must also stress that I do not intend these chapters to be a complete history of loop quantum gravity. I am sure that each of the people who worked on the theory would tell the story in a different way. The story I tell is sketchy and leaves out many episodes and steps in the theory's development. Worse, it leaves out many of the people who at one time or another have contributed something important to the theory.

The story of loop quantum gravity really begins in the 1950s with an idea that came from what would seem to be a totally different subject – the physics of superconductors. Physics is like this: the few really good ideas are passed around from field to field. The physics of materials such as metals and superconductors has been a very fertile source of ideas about how physical systems might behave. This is undoubtedly because in these fields there is a close interaction between theory and experiment which makes it possible to discover new ways for physical systems to organize themselves. Elementary particle physicists do not have access to such direct probes of the systems they model, so it has happened that on several occasions we have raided the physics of materials for new ideas.

Superconductivity is a peculiar phase that certain metals can be put into in which their electrical resistance falls to zero. A metal can be turned into a superconductor by cooling it below what is called its *critical temperature*. This critical temperature is usually very low, just a few degrees above absolute zero. At this temperature the metal undergoes a change of phase something like freezing. Of course, it is already a solid, but something profound happens to its internal structure which liberates the electrons from its atoms, and the electrons can then travel through it with no resistance. Since the early 1990s there has been an intensive quest to find materials that are superconducting at room temperature. If such a material were to be found there would be profound economic implications, as it might greatly reduce the cost of supplying electricity. But the set of ideas I want to discuss go back to the 1950s, when people first understood how simple superconductors work. A seminal step was the invention of a theory by John Bardeen, Leon Cooper and John Schrieffer, known as the BCS theory of superconductivity. Their discovery was so important that it has influenced not only many later developments in the theory of materials, but also developments in elementary particle physics and quantum gravity.

You may remember a simple experiment you did at school with a magnet, a piece of paper and some iron filings. The idea was to visualize the field of the magnet by spreading the filings on a piece of paper placed over the magnet. You would have seen a series of curved lines running from one pole of the magnet to the other (Figure 19). As your teacher may have told you, the apparent discreteness of the field lines is an illusion. In nature they are distributed continuously; they only appear to be a discrete set of lines because of the finite size of the iron

filings. However, there is a situation in which the field lines
really are discrete. If you pass a magnetic field through a
superconductor, the magnetic field breaks up into discrete
field lines, each of which carries a fundamental unit of
magnetic flux (Figure 20). Experiments show that the amount
of magnetic flux passing through a superconductor is always
an integer multiple of this fundamental unit.

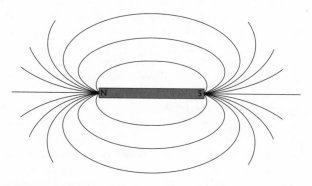

FIGURE 19
Field lines between two poles of an ordinary magnet, in air.

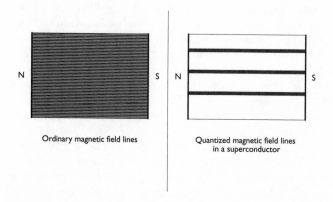

Ordinary magnetic field lines

Quantized magnetic field lines
in a superconductor

FIGURE 20
The magnetic field of a superconductor breaks up into discrete flux lines,
each carrying a certain minimum amount of the field.

This discreteness of the magnetic field lines in super-conductors is a curious phenomenon. It is unlike the discreteness of the electric charge, or of matter, in that it has to do with a field that carries a force. Furthermore, it seems that we can turn it on and off, depending on the material the magnetic field is passing through.

The electric field has field lines as well, although there is no equivalent of the iron filings experiment that allows us to see them. But in all circumstances we know about they are continuous: no material has been found which functions like an electric superconductor to break electric field lines into discrete units. But we can still imagine something like an electric superconductor, in which the field lines of the electric field would be quantized. This idea has been very successful in explaining a result from another seemingly unrelated subject: experiments show that protons and neutrons are each composed of three smaller entities called quarks.

We have good evidence that there are quarks inside protons and neutrons, just as there are electrons, protons and neutrons inside the atom. There is one difference, however, which is that the quarks seem to be trapped inside the protons. No one has ever seen a quark moving freely, that was not trapped inside a proton, neutron or other particle. It is easy to free electrons from atoms – one needs only to supply a little energy, and the electrons jump out of the atom and move freely. But no one has found a way to free a quark from a proton or neutron. We say that the quarks are *confined*. What we then need to understand is whether there is a force that can act as the electric field does in holding electrons around the nucleus, but that does so in such a way that the quarks can never come out.

From many different experiments we know that the force that holds the quarks together inside a proton is quite similar to the electric force. For one thing, we know that force is transmitted by a field that forms lines like electric and magnetic field lines. These lines connect charges which are carried by the quarks, just as electric field lines connect positive and negative electric charges. However, the force between quarks is rather more complicated than the electric force, for which there is only one kind of charge. Here there are three different varieties of charge, each of which can be positive or negative. These different charges are called *colours*, which is why the theory that describes them is called *quantum chromodynamics*, QCD for short. (This has nothing to do with ordinary colours, it is just a vivid terminology which reminds us there are three kinds of charge.) Imagine two quarks held together by some colour-electric field lines, as shown in Figure 21. Experiments show that when the two quarks are very close to each other they seem to move almost freely, as if the force between them is not very strong. But if an attempt is made to separate the two quarks, the force holding them together rises to a constant value, which does not fall off no matter how far apart they are pulled. This is very different from the electric force, which becomes weaker with increasing distance.

There is a simple way to picture what is happening. Imagine that the two quarks are connected by a length of string. This string has the peculiar property that it can be stretched however far we want. But to separate the quarks we must stretch the string, and this requires energy. No matter how long the string already is, we are going to have to put more energy in to stretch it more. To put energy into the string we must pull on it, which means that there is a force between the quarks. No matter how far apart the quarks are, to pull them farther apart you must stretch the string more, which means that there is always a force between them. As shown in Figure 21, no matter how far apart they may be they are still connected to each other by the string. This stringy picture of the force that holds quarks together is very successful, and explains the results of many experiments. But it brings with it a question: what is the string made of? Is it itself a fundamental entity, or is it composed of anything simpler? This is a question that generations of elementary particle physicists have worked to answer.

The one big clue we have is that the string stretched between two quarks behaves just like a line of magnetic flux in a superconductor. This suggests a simple hypothesis:

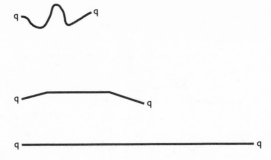

FIGURE 21

Quarks are held together by strings made of quantized flux lines of a field, called the QCD field, which are analogous to the quantized magnetic flux lines in a superconductor (Figure 20). As the quarks are pulled farther apart, the flux lines are stretched, and the force between the quarks is the same no matter how far apart they are. The result is that the quarks cannot be pulled apart.

perhaps empty space is very like a superconductor, except that what ends up discrete is the lines of force holding the colour charges of quarks together rather than the lines of magnetic flux. In this picture the lines of force between the coloured charges on the quarks are analogous to the electric rather than the magnetic field. So this hypothesis can be put very succinctly as follows: empty space is a colour-electric superconductor. This has been one of the most seminal ideas in elementary particle physics over the last few decades. It explains why quarks are confined in protons and neutrons, as well as many other facts about elementary particles. But what is really interesting is that the idea, clear as it is, contains a puzzle, for it can be looked it in two quite different ways.

One can take the colour-electric field as the fundamental entity, and then try to understand the picture of a string stretched between the quarks as a consequence of space having properties that make it something like an electric version of a superconductor. This is the route taken by those physicists who work on QCD. For them, the key problem is to understand why empty space has properties that make it behave in certain circumstances like a superconductor. This is not as crazy as it sounds. We understand that in quantum theory space must be seen to be full of oscillating random fields, as discussed in Chapter 6. So we may imagine that these vacuum fluctuations sometimes behave like the atoms in a metal in a way that leads to large-scale effects like superconductivity.

But there is another way to understand the picture of quarks held together by stretched strings. This is to see the strings themselves as fundamental entities, rather than as made up of the force lines of some field. This picture led to the original string theory. According to the first string theorists, the string is fundamental and the field is only an approximate picture of how the strings behave in some circumstances.

We thus have two pictures. In one, the strings are fundamental and the field lines are an approximate picture. In the other, the field lines are fundamental and the strings are the derived entities. Both have been studied, and both have had some success in explaining the results of experiments. But surely only one can be right? During the 1960s there was only one picture – the string picture. During this period were planted the seeds that would lead, two decades later, to the invention of string theory as a possible quantum theory of gravity. QCD was invented in the 1970s and quickly superseded the string picture as it seemed more successful as a fundamental theory. But string theory was revived in the mid-1980s, and now, as we enter the twenty-first century, both theories are thriving. It may still be that one is actually more fundamental than the other, but we have not yet been able to decide which.

There is a third possibility, which is that both the string picture and the field picture are just different ways of looking at the same thing. They would then be equally fundamental, and no experiment could decide between them. This possibility excites many theorists, as it challenges some of our deepest instincts about how to think about physics. It is called the *hypothesis of duality*.

I should emphasize that this hypothesis of duality is not the same as the wave–particle duality of quantum theory. But it is as important as that principle or the principle of relativity.

Like the principles of relativity and quantum theory, the hypothesis of duality tells us that two seemingly different phenomena are just two ways of describing the same thing. If true, it has profound implications for our understanding of physics.

The hypothesis of duality also addresses an issue that has plagued physics since the middle of the nineteenth century, that there seem to be two kinds of things in the world: particles and fields. This dualistic description seems necessary because, as we have known since the nineteenth century, charged particles do not interact directly with one another. Instead, they interact via the electric and magnetic fields. This is behind many observed phenomena, including the fact that it takes a finite speed for information to travel between particles. The reason is that the information travels via waves in the field.

Many people have been troubled by the need to postulate two very different kinds of entities to explain the world. In the nineteenth century people tried to explain fields in terms of matter. This was behind the famous aether theory, which Einstein so effectively quashed. Modern physicists try instead to explain particles in terms of fields. But this does not eliminate all the problems. Some of the most serious of these problems have to do with the fact that the theory of fields is full of infinite quantities. They arises because the strength of the electric field around a charged particle increases as one gets closer to the particle. But a particle has no size, so one can get as close as one likes to it. The result is that the field approaches infinity as one approaches the particle. This is responsible for many of the infinite expressions that arise in the equations of modern physics.

There are two ways to resolve this problem, and we shall see that both play a role in quantum gravity. One way is to deny that space is continuous, which then makes it impossible to get arbitrarily close to a particle. The other way is through the hypothesis of duality. What one can do is replace the particles by strings. This may work because from a distance one cannot really tell if something is a point or a little loop. But if the hypothesis of duality is true, then the strings and the fields may be different ways of looking at the same thing. In this way, by embracing the hypothesis of duality, several of the problems that have clouded our understanding of physics for almost two centuries may be resolved.

I personally believe in this hypothesis. To explain why, I can tell the story of two seminars I attended just before and just after I started graduate school in 1976. I happened to have my interview at Harvard on the day that Kenneth Wilson was giving a talk about QCD. Wilson is one of the most influential theoretical physicists, responsible for several innovations, including the subject of that seminar. He had come up with a remarkable way to understand the electric-superconductor picture of empty space which has since been a major influence on the life's work of many physicists, including myself.

Wilson asked us to imagine that space is not continuous, but is instead represented by a kind of graph, with points connected in a regular arrangement by lines, as shown in Figure 22. We call such a regular graph a *lattice*. He suggested to us that the distance between the points of his lattice was very small, much smaller than the diameter of a proton. So it would be hard to tell from experiment that the lattice was there at all. But conceptually it made a huge difference to think of space as a discrete lattice rather than a continuum. Wilson showed us that there was a very simple way to describe the colour-electric field of QCD by drawing field lines on his lattice. Rather than trying to show that empty space was like a superconductor, he simply assumed that the field lines were discrete entities which could move around his lattice. He wrote down simple rules to describe how they moved and interacted with one another.

Ken Wilson then argued in completely the opposite direction to everyone who had previously thought about these questions. He showed us that if there was one kind of electric charge, as in ordinary electricity, the field lines would have the tendency to group collectively in such a way that when they got very long they would lose the property of discreteness and behave like ordinary electric field lines. So he

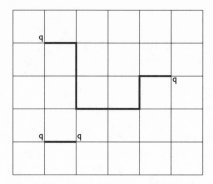

FIGURE 22

Quarks and strings as conceived of by Kenneth Wilson. Space is imagined as a lattice made of nodes connected by edges. The quarks can live only on the nodes of the lattice. The strings, or quantized tubes of flux of the field, connect the quarks but can exist only on the edges of the lattice. The distance between the nodes is assumed to be finite, but much smaller than the size of the proton. For simplicity, the lattice shown here is drawn in two dimensions only.

derived the ordinary experience of the world from his theory, rather than the reverse. But when there were three kinds of charge, as there were with quarks, then no matter how big they got, they would always stay discrete. And there would be a constant force between the quarks. The rules governing Wilson's theory were very simple – so simple, in fact, that one could explain them to a child.

Wilson's loops, as everyone has called them since, later became a major theme of my life as a theoretical physicist. I don't actually recall reflecting on the seminar afterwards, but I do recall its presentation very vividly. Nor did I then, so far as I remember, formulate the simple argument that came to me many years later: if physics is much simpler to describe under the assumption that space is discrete, rather than continuous, is not this fact itself a strong argument for space being discrete? If so, then might space look, on some very small scale, something like Wilson's lattice?

Next autumn I started graduate school, and later that year I came in one day to find a great buzz of excitement amongst the theorists. The Russian theorist Alexander Polyakov was visiting and was to give a talk that afternoon. In those days there were great schools of theoretical physics in the Soviet Union, but their members were seldom allowed to travel to the West. Polyakov was the most creative and most charismatic of them, and we all went along to his seminar. I recall someone with disarming warmth and informality, under which there was hidden (but not too well) someone with unlimited confidence.

He began by telling us that he had dedicated his life to pursuing a foolish and quixotic vision, which was to re-express QCD in a form in which the theory could be solved exactly. His idea for doing this was to recast QCD completely as a theory of the dynamics of lines and loops of colour-electric flux. These were the same as Wilson's loops, and indeed Polyakov had independently invented the picture of QCD on a discrete lattice. But in this seminar at least he worked without the lattice, to try to pull out from the theory a description in which the quantized loops of electric flux would be the fundamental entities. A physicist working without a lattice is something like a trapeze artist working without a net. There is an ever present danger that a false move will lead to a fatal result. In physics the fatalities arise from confrontations with infinite and absurd mathematical expressions. As we mentioned earlier, such expressions arise in all quantum theories based on continuous space and time. In his seminar Polyakov showed that despite these infinities, one could give physical meaning to loops of electric flux. If he did not succeed completely in solving the resulting equations, his seminar was altogether an assertion of faith in the hypothesis of duality – that the strings are as fundamental as the electric field lines.

The idea of duality is still a major driving force behind research in elementary particle physics and string theory. Duality is the very simple view that there are two ways of looking at the same thing – either in terms of strings or in terms of fields. But so far no one has been able to show that duality is applicable in ordinary QCD. It has been shown to be valid in very specialized theories which depend on very specific simplifying assumptions. Either the dimensionality of space is reduced from three to one, or a great deal of additional symmetry is added, which leads to a theory that can be understood much more easily. But even if it has not yet solved the problem that inspired its invention, duality has turned out to be a central concept in quantum gravity. How this happened is a very typical tale of how good scientific ideas can spread far from their point of origin, for I rather doubt that either Wilson or Polyakov originally considered how their idea might be applied to a quantum theory of gravity.

Like many good ideas, this one needed several goes to get it right. Inspired by what I had heard from Wilson and Polyakov, and further lessons on lattice theories I got from Gerard 't Hooft, Michael Peskin and Stephen Shenker during my first year of graduate school, I set out to formulate quantum gravity in terms of Wilson's lattice theory. Using some ideas borrowed from several people, I was able to concoct such a theory, which enabled me to spend a year or so learning the various techniques developed by Polyakov, Wilson and others by applying them to my version of quantum gravity. I wrote up and sent out a long paper about it and waited for a reaction. As was common in those days, the only response was a stack of postcards from far-away places requesting copies of the paper. There was of course the inevitable request from the U.S. Army research lab, which reminded us that someone somewhere was being paid to think about the possible military applications of whatever young graduate students were up to. It is strange to recall those days, not so long ago, when we typed our papers on IBM Selectric typewriters, got a professional in the basement to draw the illustrations, and then stuffed the copies individually into envelopes and mailed them out. These days we write our papers on laptops and upload them to electronic archives from where they are immediately available on the Internet. I doubt many of our current students have seen either an IBM Selectric typewriter or a postcard requesting a preprint. Many have never even gone to the library to read a paper printed in a journal.

A few months later I realized that the paper was basically wrong. It was a brave attempt, but fatally flawed. Still, it got me a few invitations to conferences. I don't think Stephen Hawking was very happy when I used the occasion of his invitation to give a talk at a conference he organized to explain why making a lattice theory of gravity was not a very smart thing to do. Some people seemed to like the idea, but I did not see what else I could do – it was a bad idea, and I had the responsibility to explain why.

At another conference I left a copy of the paper in the mailbox of someone called Ashok Das, who had told me he was interested in doing something similar. Bryce DeWitt, who is justly thought of as the father of serious research in quantum gravity, looked for his mail in the same box and assumed that my paper was intended for him. I'm sure he saw all its shortcomings, but he was still kind enough to ask me to join him as a postdoc. I owe my career to Bryce's mistake. At that time I was being told that I had committed professional suicide by working on quantum gravity and that I was unlikely to get any job at all.

What was wrong with my first paper was that Wilson's lattice was an absolute, fixed structure and thus clashed with the relational nature of Einstein's theory of gravity. So my theory did not contain gravity and had nothing at all to do with relativity. To fix this, the lattice itself would have to become a dynamical structure which could evolve in time. The key lesson I learned from this failed attempt was that one cannot fashion a successful quantum theory of gravity out of objects moving against a fixed background.

At about this time I met Julian Barbour, a physicist and philosopher who lives in a little village near Oxford. Julian had left the academic world after his Ph.D. in order to have the freedom to think deeply about the nature of space and time. He supported himself by translating Russian scientific journals into English and, away from the usual pressures of academic life, he used his considerable linguistic skills to read deeply into the history of our understanding of space and time. He had understood from his study the importance of the idea that space and time are relational, and he had then applied this wisdom to modern physics. He was I believe the first person to gain a deep understanding of the role this idea plays in the mathematical structure of Einstein's theory of relativity. In a series of papers, first alone and then with an Italian friend, Bruno Bertotti, he showed how to formulate mathematically a theory in which space and time were nothing but aspects of relationships. Had Leibniz or anyone else done this before the twentieth century, it would have changed the course of science.

As it happened, general relativity already existed, but – and this is a strange thing to say – it was widely misunderstood, even by many of the physicists who specialized in its study. Unfortunately, general relativity was commonly regarded as a machine that produces spacetime geometries, which are then to be treated as Newton treated his absolute space and time: as fixed and absolute entities within which things move. The question then to be answered was which of these absolute spacetimes describes the universe. The only difference between this and Newton's absolute space and time is that there is no choice in Newton's theory, while general relativity offers a selection of possible spacetimes. This is how the theory is presented in some textbooks, and there are even some philosophers, who should know better, who seem to interpret it this way. Julian Barbour's important contribution was to show convincingly that this was not at all the right way to understand the theory. Instead, the theory has to be understood as describing a dynamically evolving network of relationships.

Julian was of course not the only person to learn to see general relativity in this way. John Stachel also came to this understanding, at least partly through his work as the first head of the project to prepare Einstein's collected papers for publication. But Julian came to the study of general relativity equipped with a tool that no one else had – the general mathematical formulation of a theory in which space and time are nothing but dynamically evolving relationships. Julian was then able to show how Einstein's theory of general relativity could be understood as an example of just such a theory. This demonstration laid bare the relational nature of the description of space and time in general relativity.

Since then, Julian Barbour has become known to most people working in relativity, and recently he has become even more widely known and appreciated as a result of the publication of his radical theories on the nature of time. But in the early 1980s few people knew of his work, and I was very fortunate to meet him shortly after I had realized that my lattice gravity theory was in trouble. During this meeting he explained to me the meaning of space and time in general relativity, and the role of the relational concept in it. This gave me the conceptual language to understand why my calculations were showing that gravity was nowhere to be found in the theory I had constructed. What I needed to do was invent something like Wilson's lattice theory, but in which there was no fixed lattice, so that all the structures were dynamical and relational. A set of points connected by edges – in other words a *graph* – is a good example of a system defined by relationships. But what I had done wrong was to base the theory on a fixed graph. Instead, the theory should produce the graph, and it should not mirror any pre-existing geometry or structure. It should rather evolve according to rules as simple as those that Wilson had given for the motion of loops on his lattice. It was to be ten years before a way appeared which made this possible.

During those ten years I spent my time on a variety of unsuccessful attempts to apply techniques from particle physics to the problem. These techniques were all background dependent, in that they assumed that you could fix a single classical spacetime geometry and study how quantized gravitational waves, called gravitons, move and interact on the background. We tried lots of different approaches, but they all failed. Besides this I wrote a few papers on supergravity, the new theory of gravity which had been invented by one of my advisors, Stanley Deser, and others. Those attempts also came to nothing. Then I wrote a few papers about the implications of the entropy of black holes, making various speculations about their connection with problems in the foundations of quantum mechanics. Looking at them now, it seems to me that these papers were the only interesting things I did during those years, but I have no evidence that very many people ever read them. Certainly there was no interest in, and no market for, young people applying ideas from quantum black holes to fundamental issues in quantum theory.

Looking back, I am quite puzzled about why I continued to have a career. One sure reason was because at that time very few people worked on quantum gravity, so there was little competition. I was not actually getting anywhere, but people seemed to be interested in at least that part of my work in which I tried to apply techniques from particle physics to quantum gravity, even if what I had to report was that these were not very smart things to do. No one else was getting very far either, so there was room for the kind of people who prefer trying new things to following the research programs of older people, or who thrive on stealing ideas from one field and applying them in another. I doubt very much that I would have a career in the present-day environment, which is much more competitive, and in which the jobs are controlled by older people who feel confident that they are working on the right approach to quantum gravity. This allows them – but I should really say 'us', for I am now one of the older people who hires postdocs – to feel justified in using the enthusiasm a young researcher shows for our own research program as a measure of that researcher's promise.

For me, as for many people working in this area, the turning point came with the revival of string theory as a possible quantum theory of gravity. I shall come to string theory in the next chapter. For now I shall say only that, having experienced the invention and failure of a whole series of wrong approaches to quantum gravity, I, along with many other physicists, was quite optimistic about what string theory could do for us. At the same time, I was also completely convinced that no theory could succeed if it was based on things moving in a fixed background spacetime. And however successful string theory was at solving certain problems, it was still very much a theory of this kind. It differed from a conventional theory only inasmuch as the objects moving in the background were strings rather than particles or fields. So it was clear right away, to me and to a few others, that while string theory might be an important step towards a quantum theory of gravity, it could not be the complete theory. But nevertheless, as it did for many other physicists, string theory changed the direction of my research. I began to look for a way to make a background independent theory which would reduce to string theory as an approximation useful in situations in which spacetime could be regarded as a fixed background.

To get inspiration for this project I recalled the seminar given by Polyakov that had so excited me as a beginning graduate student. I wondered whether I could use the method he had used, which was to try to express QCD in terms in which the fundamental objects were loops of colour-electric flux. I needed a theory in which there was no lattice to get in the way, and he had worked without a lattice. I worked for about a year on this idea, with Louis Crane. I was then a postdoc at the University of Chicago, and Louis Crane was a graduate student. He was older than me, but he had actually been a child prodigy, perhaps the last of a distinguished series of scientists and scholars that the University of Chicago admitted to college in their early teens. He had suffered the misfortune of being expelled from graduate school for leading a strike against the invasion of Cambodia, and it had taken him ten years to find his way back to graduate school. Louis has since become one of a handful of mathematicians who has made significant creative contributions to the development of our ideas about quantum gravity. Some of his contributions have been absolutely seminal for developments in the field. I was very fortunate to become his friend at that time, and count myself lucky to be his friend still.

Louis and I worked on two projects. In the first we tried to formulate a gravitational theory based on the dynamics of interacting loops of quantized electric flux. We failed to formulate a string theory, and as a result we published none of this work, but it was to have very important consequences. In the second project we showed that a theory in which spacetime was discrete on small scales could solve many of the problems of quantum gravity. We did this by studying the implications of the hypothesis that the structure of spacetime was like a fractal at Planck scales. This overcame many of the difficulties of quantum gravity, by eliminating the infinities and making the theory finite. We realized during that work that one way of making such a fractal spacetime is to build it up from a network of interacting loops. Both collaborations with Louis Crane persuaded me that we should try to construct a theory of spacetime based on relationships among an evolving network of loops. The problem was, how should we go about this?

This was how things stood when a discovery was made that completely changed how we understand Einstein's theory of general relativity.

··

CHAPTER 10

KNOTS, LINKS AND KINKS

During the year I was working with Louis, a young postdoc named Amitaba Sen published two papers which excited and mystified many people. We read them with a great deal of interest, for what Sen was doing was attempting to make a quantum theory from supergravity. Embedded in the papers were several remarkable formulas in which Einstein's theory of gravity was expressed in a much simpler and more beautiful set of equations than Einstein had used. Several of us spent many hours discussing what would happen if we could somehow find a way to base quantum gravity on this much simpler formulation. But none of us did anything at the time.

The one person who took Sen's equations seriously was Abhay Ashtekar. Abhay had been trained as a classical relativist, and early in his career had done important work in that area, but more recently he had set his sights on a quantum theory of gravity. Being mathematically inclined, Ashtekar saw that Sen's equations contained the core of a complete reformulation of Einstein's general theory of relativity, and by a year later he had done just that: fashioned a new formulation of general relativity. This did two things: it vastly simplified the mathematics of the theory, and expressed it in a mathematical language which was very close to that used in QCD. This was exactly what was needed to transform quantum gravity into a real subject, one in which it would in time become possible to do calculations that yielded definite predictions about the structure of space and time on the Planck scale.

I invited Abhay to give a talk about this at Yale, where I had just become an assistant professor. At the talk was a graduate student named Paul Renteln, from Harvard, who had also been studying Sen's papers. It was clear to us that Ashtekar's formulation would be the key to further progress. Afterwards, I drove Abhay to the airport in Hartford. On the hour's drive between New Haven and Hartford my car had not one but two flat tyres – and Abhay still just caught his flight. He had to hitch a ride for the last few miles, while I waited on the side of the road for help.

When I finally got home I sat down immediately and began to apply to the new formalism of Sen and Ashtekar the methods Louis Crane and I had developed during our unsuccessful attempts to re-invent string theory. A few weeks later there began a semester-long workshop in quantum gravity at the Institute for Theoretical Physics in Santa Barbara. By another piece of luck, I had convinced the authorities at Yale to let me spend a semester there, just after they had hired me. As soon as I got there I recruited two friends, Ted Jacobson and Paul Renteln. We found right away that a very simple picture of the quantum structure of space emerged if we used something very like the electric super-conductor picture for the flux lines of the gravitational field. At first I worked with Paul. Fearful of the infinities that come with continuous space, we used a lattice, much like Wilson's lattice. We found that the new form of the Einstein equations implied very simple rules for how the loops interact on the lattice. But we ran into the same problem as I had ten years before: how to get rid of the background imposed by the use of a fixed lattice.

Ted Jacobson suggested that we try to follow Polyakov and work without a lattice. I have already described the result, in Chapter 3. The next day we were standing in front of a blackboard, staring at something which it had never occurred to us, nor anyone else, to even look for. These were exact solutions to the full equations of the quantum theory of gravity.

What we had done was to apply the usual methods for constructing a quantum theory to the simple form of the equations for general relativity that Sen and Ashtekar had discovered. These led to the equations for the quantum theory of gravity. These equations had first been written down in the 1960s, by Bryce DeWitt and John Wheeler, but we found new forms for them which were dramatically simpler. We had to plug into these equations formulas that described possible quantum states of the geometry of space and time. On an impulse I tried something that Louis Crane and I had played with, which was to build these states directly from the expressions Polyakov used to describe the quantized loops of electric fields. What we found was that, as long as the loops did not intersect, they satisfied the equations. They looked like the loops in Figure 23.

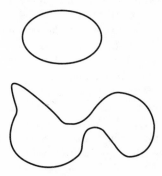

FIGURE 23

Quantum states of the geometry of space are expressed in loop quantum gravity in terms of loops. These states are exact solutions to the equations of quantum gravity, as long as there are no intersections or kinks in the loops.

It took us a few days of hard work to find still more solutions. We found that even if the loops intersected, we could still combine them to make solutions provided certain simple rules were obeyed. In fact, we could write down an infinite number of these states – all we had to do was draw loops and apply some simple rules whenever they intersected.

It took many years for us and others to work out the implications of what we had found in those few days. But even at the start we knew that we had in our grasp a quantum theory of gravity that could do what no theory before it had done – it gave us an exact description of the physics of the Planck scale in which space is constructed from nothing but the relationships among a set of discrete elementary objects. These objects were still Wilson's and Polyakov's loops, but they no longer lived on a lattice, or even in space. Instead, their interrelations defined space.

There was one step to go to complete the picture. We had to prove that our solutions really were independent of the background space. This required us to show that they solved an additional set of equations, known as diffeomorphism constraints, which expressed the independence of the theory from the background. These were supposed to be the easy equations of the theory. Paradoxically, the equations we had solved so easily, the so-called Wheeler–DeWitt equations, were supposed to be the hard ones. At first I was very optimistic, but it turned out to be impossible to invent quantum states that solved both sets of equations. It was easy to solve one or the other, but not both.

Back at Yale the next year, we spent many fruitless hours with Louis Crane trying to do this. We pretty much convinced ourselves it was impossible. This was very frustrating because it was easy to see what the result would be: if we could only get rid of the background, we would have a theory of nothing but loops and their topological relationships. It would not matter where in space the loops were, because the points in space would have no intrinsic meaning. What would matter would be how the loops intersected one another. It would also matter how they knotted and linked.

I realized this one day while I was sitting in my garden in Santa Barbara. Quantum gravity would be reduced to a theory of the intersecting, knotting and linking of loops. These would give us a description of quantum geometry on the Planck scale. From the work I had done with Paul and Ted, I also knew that the quantum versions of the Einstein equations we had invented could change the way the loops linked and knotted with one another. So the relationships among the loops could change dynamically. I had thought about intersecting loops, but I had never wondered about how loops could knot or link.

I went inside and called Louis Crane. I asked him whether mathematicians knew anything about how loops might knot and link. He said, yes, there is a whole field devoted to the subject, called knot theory. He reminded me that I had had dinner a few times in Chicago with one of its leading thinkers, Louis Kauffman. So the last step was to rid the theory of any dependence on where the loops were in space. This would reduce our theory to the study of knots, links and kinks, as James Hartle, one of the leading American relativists, teasingly began to call it shortly afterwards. But this was not so easy, and we were not able to take this step for over a year. We tried very hard, with Louis and others, but we could not do it.

The workshop at Santa Barbara had closed with a conference at which our new results were first presented. There I had met a young Italian scientist, Carlo Rovelli, who had just been awarded his Ph.D. We didn't talk much, but shortly after he wrote to ask if he could come to visit us at Yale. He arrived that October, and took a room in Louis Crane's apartment. The first day he was there I explained to him that there was nothing to do, because we were completely stuck. The work had looked promising, but Louis and I had found the last step to be impossible. I told Carlo he was welcome to stay, but perhaps given the sad state of the subject he might prefer to go back to Italy. There was an awkward moment. Then, looking for something to talk about, I asked him if he liked to sail. He replied that he was an avid sailor, so we abandoned science for the day and went straight to the harbour where the Yale sailing team kept its boats, and took out a sailing dinghy. We spent the rest of the afternoon talking about our girlfriends.

I didn't see Carlo the next day. The day after that he appeared at the door to my office and said, 'I've found the answer to all the problems.' His idea was to make one more reformulation of the theory, so that the basic variables were nothing but the loops. The problem was that the theory up till then depended both on the loops and on the field flowing around the loop. Carlo saw that it was the dependence on the field that was making it impossible to proceed. He also saw how to get rid of it, by using an approach to quantum theory invented by his mentor, Chris Isham, at Imperial College. Carlo had found that applying it to the loops gave exactly what we needed. It took us no more than a day to sketch the whole picture. In the end we had a theory of the kind that Polyakov had spoken about as his great dream: a theory of pure loops which described an aspect of the real world in equations so simple they could be solved exactly. And when it was used to construct the quantum version of Einstein's theory of gravity, the theory depended only on the relationships of the loops to one another – on how they knot, link and kink. Within days we had shown that one can construct an infinite number of solutions to all the equations of quantum gravity. For example, there is one solution for every possible way to tie a knot.

A few weeks later we went to Syracuse University, which by then was the centre of work stemming from Ashtekar's and Sen's discoveries, and Carlo gave the first seminar on the new theory of quantum gravity. On the way to the airport we were rear-ended by a guy in a very flashy car. No one was hurt, and the rear bumper of my old Dodge Dart was barely scratched, but his Maserati was wrecked. Still, we made it. The next day Carlo had a high fever, but he got through the seminar, and at the end there was a long, appreciative silence. Abhay Ashtekar said it was the first time he had seen something that might be the quantum theory of gravity. A few weeks after that I gave the second seminar on the new theory, in London, in front of Chris Isham, on my way to a conference in India.

In India, two ancient cultures met when I introduced the conference organizer to Carlo, who had decided impulsively to jump on a plane and come, though he had no invitation. The distinguished gentleman looked at his long hair and the sandals and clothes he had picked up while wandering alone for two days through the back streets of Bombay, and sputtered, 'Mr Rovelli, but didn't you get my letter saying the meeting is closed?' Carlo smiled and replied, 'No, but didn't you get mine?' He was given the best room in the hotel, and Air India put him in first class on his flight home to Rome.

Thus was born what is now called loop quantum gravity. It took several years of work, first with Carlo and then as part of a growing community of friends and colleagues, to unravel the meaning of the solutions to the quantum gravity equations we had found. One straightforward consequence is that quantum geometry is indeed discrete. Everything we had done had been based on the idea of a discrete line of force, as in a magnetic field in a superconductor. Translated into the loop picture of the gravitational field, this turns out to imply that the area of any surface comes in discrete multiples of simple units. The smallest of these units is about the Planck area, which is the square of the Planck length. This means that all surfaces are discrete, made of parts each of which carries a finite amount of area. The same is true of volume.

To arrive at these results we had to find a way to eliminate the infinities that plague all expressions in quantum theories of fields. I had an intuition, stemming from my past conversations with Julian Barbour, and the work I had done with Louis Crane, that the theory should have no infinities. Many physicists have speculated that the infinities come from some mistaken assumption about the structure of space and time on the Planck scale. From the older work it was clear to me that the wrong assumption was the idea that the geometry of spacetime was fixed and non-dynamical. When calculating the measures of geometry, such as area and volume, one had to do it in just the right way to eliminate any possible contamination from non-dynamical, fixed structures. Exactly how to do this was a technical exercise that cannot be explained here. But in the end it did turn out that as long as one asks a physically meaningful question, there will be no infinities.

In my experience it really is true that as a scientist one has only a few good ideas. They are few and far between, and come only after many years of preparation. What is worse, having had a good idea one is condemned to years of hard work developing it. The idea that area and volume would be discrete had come to me in a flash as I was trying to calculate the volume of some quantum geometry, while I was sitting for an hour in a noisy room in a garage waiting for my car to be fixed. The page of my notebook was filled with many messy integrals, but all of a sudden I saw emerge a formula for counting. I had begun to calculate a quantity on the assumption that the result was a real number, but found instead that, in certain units, all the possible answers would be integers. This meant that areas and volumes cannot take any value, but come in multiples of fixed units. These units correspond to the smallest areas and volumes that can exist. I showed these calculations to Carlo, and a few months later, during a period we spent working together at the University of Trento, in the mountains of north-east Italy, he invented an argument that showed that the basic unit of area could not be taken to zero. This meant there was no way to avoid the conclusion that if our theory were true, space had an 'atomic' structure.

I well remember our work in Trento for another reason. In the previous year one of our students, Bernd Bruegmann, had come to my office with a very disturbed look on his face. His thesis problem was to apply the new methods from loop quantum gravity to QCD on a lattice, and see whether the properties of protons and neutrons would emerge. While doing so he did what good scientists should do, but which we had not, which was to check the literature thoroughly. He had found a paper in which methods very similar to ours had already been applied to QCD by two people we had never heard of, Rodolfo Gambini and Anthony Trias, who were working in Montevideo and Barcelona.

Scientists are human, and we all suffer from the need to feel that what we do is important. Pretty much the worst thing that can happen to a scientist is to find that someone has made the same discovery before you. The only thing worse is when someone publishes the same discovery you made, after you've published it yourself, and does not give you adequate credit. It was true that we had discovered the method of working with loops in the realm of quantum gravity rather than in QCD, but there was no avoiding the fact that the method we had developed was quite close to the one that Gambini and

Trias had already been using for several years in their work on QCD. Even though they had been publishing in the *Physical Review*, which is a major journal, we had somehow missed seeing their work.

With a heavy heart we did the only thing we could, which was to sit down and write them a very apologetic letter. We heard nothing from them until one afternoon in Trento, when Carlo got a phone call from Barcelona. Our letter had finally reached them. They had tracked us down to Trento, and asked if we would still be there tomorrow. The next morning they arrived, having driven most of the night across France and northern Italy. We spent a wonderful day showing each other our work, which was thankfully complementary. They had applied the method to QCD, while we had applied it to quantum gravity. Anthony Trias did most of the talking, while Rodolfo Gambini sat at the back of the room and at first hardly said anything. But we soon found that Rodolfo was a creative scientist of the first order. Just how creative we learned over the next few months, as he quickly invented a new approach to doing calculations in loop quantum gravity.

Since then Gambini has been one of the leaders in the field of quantum gravity, often working in collaboration with Jorge Pullin at Penn State University and a very good group of young people he trained in Montevideo. They have discovered many more solutions to the equations of quantum gravity, and resolved several important problems that came up along the way.

It also must be said that, despite his quiet nature, Rodolfo Gambini has been more or less single-handedly responsible for reviving physics in both Venezuela and Uruguay after its total destruction by the military dictatorships. Just what this meant was brought home to me the first time I visited Montevideo. It was the middle of winter, and we did physics with Rodolfo and his group in a run-down old convent, without heat or computers, fighting off the cold by drinking a continuous supply of matte (a kind of tea) that was kept hot over a Bunsen burner. Now the science departments at the University of Uruguay are housed in modern buildings and facilities, built with funds that Rodolfo raised in his spare time, while keeping up a continuous flow of new ideas and calculations.

One of the most beautiful results to have come from loop quantum gravity was the discovery that the loop states could be arranged in very beautiful pictures, which are called *spin networks*. These had actually been invented thirty years earlier by Roger Penrose. Penrose had also been inspired by the idea that space must be purely relational. Going directly to the heart of the matter, as is his nature, he had skipped the step of trying to derive a picture of relational space from some existing theory, as we had. Instead, having more courage, he had sought the simplest possible relational structure that might be the basis of a quantum theory of geometry. Spin networks were what he came up with. A spin network is simply a graph, such as those shown in Figures 24 to 27, whose edges are labelled by integers. These integers come from the values that the angular momentum of a particle are allowed to have in quantum theory, which are equal to an integer times half of Planck's constant.

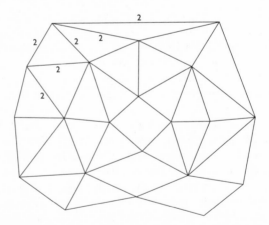

FIGURE 24

A spin network, as invented by Roger Penrose, also represents a quantum state of the geometry of space. It consists of a graph, together with integers on the edges. Only a few of the numbers are shown here.

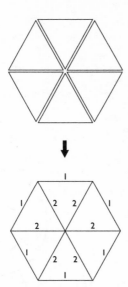

FIGURE 25
A spin network can be made by combining loops.

I had known for a long time that Penrose's spin networks should come into loop quantum gravity, but I had been afraid of working with them. When Penrose described them in his talks they always seemed so intricate that only he would be able to work with them without making mistakes. To do a calculation Penrose's way, one has to add up long series of numbers which are each either +1, 0 or −1. If you miss one sign, you're dead. Still, during a visit to Cambridge in 1994 I met Roger and asked him to tell me how to calculate with his spin networks. We did one calculation together, and I thought I had the hang of it. That was enough to convince me that spin networks would make it possible to calculate aspects of quantum geometry such as the smallest possible volume. I then showed what I had learned to Carlo, and we spent the rest of that summer translating our theory into the language of Penrose's spin networks.

When we did this we found that each spin network gives a possible quantum state for the geometry of space. The integers

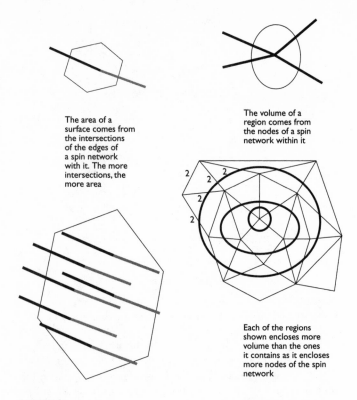

The area of a surface comes from the intersections of the edges of a spin network with it. The more intersections, the more area

The volume of a region comes from the nodes of a spin network within it

Each of the regions shown encloses more volume than the ones it contains as it encloses more nodes of the spin network

FIGURE 26

The quantization of space as predicted by loop quantum gravity. The edges of spin networks carry discrete units of area. The area of a surface comes from the intersection of one edge of a spin network with it. The smallest possible area comes from one intersection, and is about 10^{-66} of a square centimetre. The nodes of the spin networks carry discrete units of volume. The smallest possible volume comes from one node, and is about 10^{-99} of a cubic centimetre.

on each edge of a network correspond to units of area carried by that edge. Rather than carrying a certain amount of electric or magnetic flux, the lines of a spin network carry units of area. The nodes of the spin networks also have a simple meaning: they correspond to quantized units of volume. The

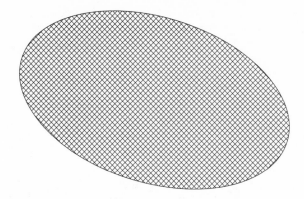

FIGURE 27

A very large spin network can represent a quantum geometry that looks smooth and continuous when viewed on a scale much larger than the Planck length. We say that the classical geometry of space is woven by making it out of a very large and complex spin network. In the spin network picture, space only seems continuous – it is actually made up of building blocks which are the nodes and edges of the spin network.

volume contained in a simple spin network, when measured in Planck units, is basically equal to the number of nodes of the network. It took much work and heartache to clarify this picture. Penrose's method was invaluable but, as I had expected, it was not easy to work with. Along the way we learned the truth of something I once heard Richard Feynman say, which is that a good scientist is someone who works hard enough to make every possible mistake before coming to the right answer.

Probably my worst moment in science came at a conference in Warsaw when a young physicist named Renate Loll, who had also been a student of Chris Isham in London, announced at the end of her talk that our calculation for the smallest possible volume was wrong. After a lot of argument it turned out that she was right, and we traced our error back to a single sign mistake. But remarkably, our basic pictures and results held up. They have since been confirmed by mathematical physicists who showed that the results we found are under-pinned by rigorous mathematical theorems. Their work tells us that the spin network picture of quantum geometry is not just a product of someone's imagination – rather, it follows directly from combining the basic principles of quantum theory with those of relativity.

The loop approach to quantum gravity is now a thriving field of research. Many of the older ideas, such as supergravity and the study of quantum black holes, have been incorporated into it. Connections have been discovered to other approaches to quantum gravity, such as Alain Connes's non-commutative approach to geometry, Roger Penrose's twistor theory and string theory.

One lesson we have learned from this experience is the extent to which science progresses quickly when people with different backgrounds and educations join forces to push back the frontiers. The relationship between theoretical physicists and mathematical physicists is not always smooth. It is rather like the relationship between the scouts who first explore the frontier, and the farmers who come after them and fence the land and make it productive. The mathematical farmers need to tie everything down, and determine the exact boundaries of an idea or a result, while we physicist scouts like our ideas when they are still a bit wild and untamed. Each has a tendency to think that they did the essential part of the work. But something we and the string theorists have both learned is that in spite of their different ways of working and thinking, it is essential that mathematicians and physicists learn to communicate and work with one another. As happened with general relativity, quantum gravity requires new mathematics as much as it requires new concepts, ideas and ways of doing calculations. If we have made real progress it is because we have discovered that people can work together to invent something that no one person could have come up with alone.

In the end, what is most satisfying about the picture of space given by loop quantum gravity is that it is completely relational. The spin networks do not live in space; their structure generates space. And they are nothing but a structure of relations, governed by how the edges are tied together at the nodes. Also coded in are rules about how the edges may knot and link with one another. It is also very satisfying that there is a complete correspondence between the classical and quantum pictures of geometry. In classical geometry the volumes of regions and the areas of the surfaces depend on the values of gravitational fields. They are coded in certain complicated collections of mathematical functions, known collectively as the metric tensor. On the other hand, in the quantum picture the geometry is coded in the choice of a spin network. These spin networks correspond to the classical description in that, given any classical geometry, one can find a spin network which describes, to some level of approximation, the same geometry (Figure 27).

In classical general relativity the geometry of space evolves in time. For example, when a gravitational wave passes a surface, the area of that surface will oscillate in time. There is an equivalent quantum picture in which the structures of the spin networks may evolve in time in response to the passage of a gravitational wave. Figure 28 shows some of the simple steps by which a spin network evolves in time. If we let a spin network evolve, we get a discrete spacetime structure. The events of this discrete spacetime are the processes by which changes of the form shown in Figure 28 occur. We can draw pictures of evolving spin networks; they look like Figures 29 to 31. An evolving spin network is very like a spacetime, but it is discrete rather than continuous. We can say what the causal relations are among the events, so it has light cones. But it also has more, for we can draw slices through it that correspond to moments of time. As in relativity theory, there are many different ways of slicing an evolving spin network, so as to see it as a succession of states evolving in time. Thus, the picture of spacetime given by loop quantum gravity agrees with the fundamental principle that in the theory of relativity there are no things, only processes.

John Wheeler used to say that on the Planck scale spacetime would no longer be smooth, but would resemble a foam, which he called spacetime foam. In tribute to Wheeler, the mathematician John Baez has suggested that evolving spin networks be called *spin foam*. The study of spin foam has sprung up since the mid-1990s. There are several different versions presently under study, invented by Mike Reisenberger, by Louis Crane

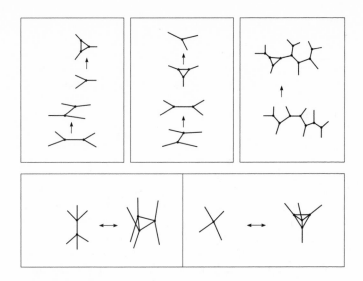

FIGURE 28

Simple stages by which a spin network can evolve in time. Each one is a quantum transition of the geometry of space. These are the quantum theoretic analogues of the Einstein equations. [From F. Markopoulou, 'Dual formulation of spin network evolution', gr-qc/9704013. All the papers referenced here as gr-qc/xxxx are available at xxx.lanl.gov.]

and John Barrett, and by Fotini Markopoulou-Kalamara. Carlo Rovelli, John Baez, Renate Loll and many of the other people who contributed to loop quantum gravity are now engaged in the study of spin foam. So this is presently a very lively area of research. Figure 32 shows a computer simulation of a world with one space and one time dimension, modelled upon ideas from spin foam theory. This is work of Jan Ambjørn, Kostas Anagnastopoulos and Renate Loll. These universes are very small, each edge corresponding to one Planck length. They do not always evolve smoothly; instead, from time to time the size of the universe jumps suddenly. These are the quantum fluctuations of the geometry. After many years, we have here a real quantum theory of the geometry of spacetime.

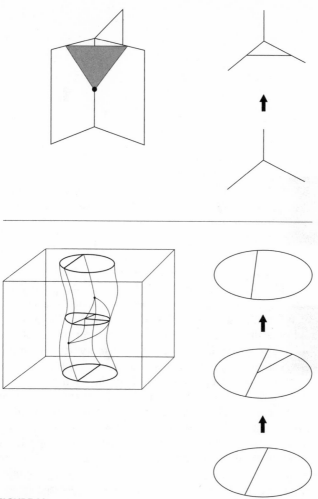

FIGURE 29

Two pictures of quantum spacetimes. Each event in the quantum spacetime is a simple change in the quantum geometry of space, corresponding to one of the moves shown in Figure 28. According to loop quantum gravity, this is what spacetime looks like if we examine it on a time scale of 10^{-43} of a second and a length scale of 10^{-33} of a centimetre. The upper picture shows a single elementary move. The lower one shows a combination of two elementary moves. [From C. Rovelli, 'The projector on physical states in loop quantum gravity', gr-qc/9806121.]

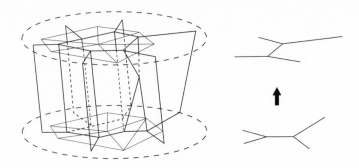

FIGURE 30
Another of the elementary moves for quantum transitions among spin networks, together with the spacetime picture that represents it. [From R. de Pietri, 'Canonical loop quantum gravity and spin foam models', gr-qc/ 9903076.]

Is the theory right? We do not know yet. In the end, it will be decided by experiments designed to test the predictions the theory makes about the discreteness of area and volume and other measures of spacetime geometry. I do want to emphasize that although it follows directly from the combination of the principles of general relativity and quantum theory, loop quantum gravity does not need to be the complete story to be true. In particular, the main predictions of the theory, such as the quantization of area and volume, do not depend on many details being correct, only on the most general assumptions drawn from quantum theory and relativity. The predictions do not constrain what else there can be in the world, how many dimensions there are or what the fundamental symmetries are. In particular, they are completely consistent with the basic features of string theory, including the existence of extra dimensions and supersymmetry. I know of no reason to doubt their truth.

Of course, in the end experiment must decide. But can we really hope for experimental confirmation of the structure of space on the Planck scale, 20 orders of magnitude smaller than the proton? Until very recently most of us were sceptical about whether we might see such tests in our lifetime. But

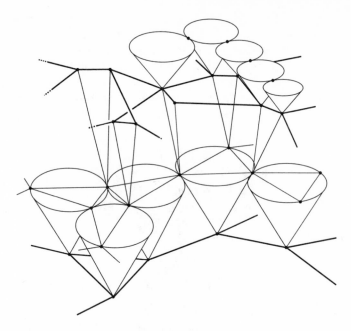

FIGURE 31

Another picture of a quantum spacetime, showing the causal futures of the events where the spin networks change. These are drawn as light cones, as in Chapter 4. [From F. Markopoulou and L. Smolin, 'The causal evolution of spin networks', gr-qc/9702025.]

now we know we were being too pessimistic. A very imaginative young Italian physicist, Giovanni Amelino-Camelia, has pointed out that there is a way to test the predictions that the geometry of space is discrete on the Planck scale. His method uses the whole universe as an instrument.

When a photon travels through a discrete geometry it will suffer small deviations from the path that classical physics predicts for it. These deviations are caused by the interference effects that arise when the photon's associated wave is scattered by the discrete nodes of the quantum geometry. For photons that we can detect, these effects are very, very small. What no one before Amelino-Camelia had thought of, though, is that the effects accumulate when a photon travels very long distances. And we can detect photons that have travelled across large fractions of the observable universe. He proposes that by carefully studying images taken by satellites of very violent events such as those that create X-ray and gamma-ray bursts, it may be possible to discover experimentally the discrete structure of space.

FIGURE 32

Computer model of a quantum spacetime, showing a universe with one space and one time dimension. The structures shown exist on scales of 10^{-33} of a centimetre and 10^{-43} of a second. We see that the quantum geometry fluctuates very strongly because of the uncertainty principle. As with the position of an electron in an atom, for such small universes the quantum fluctuations in the size of the universe are very important as they are as large as the universe itself. [These simulations are the work of Jan Ambøjrn, Kostas Anagnastopoulos and Renate Loll. They can be seen at their Web page, http://www.nbi.dk/~konstant/homepage/lqg2/.]

If these experiments do show that space has an atomic structure on the Planck scale, it will surely be one of the most exciting discoveries of early twenty-first century science. By developing these new methods we may be able to look at pictures of the discrete structure of space, just as we are now able to study pictures of arrays of atoms. And if the work I have described in the last two chapters is not completely irrelevant, what we shall see are Wilson's and Polyakov's loops, organized into Penrose's spin networks.

CHAPTER 11

THE SOUND OF SPACE IS A STRING

I am convinced that the hardest thing about doing science is not that it sometimes demands a certain level of skill and intelligence. Skills can be learned, and as for intelligence, none of us is really smart enough to get anywhere on our own. All of us, even the most independent, manage to carry our work through to completion because we are part of a community of committed and honest people. When we are stuck, most of us look for a way out in the work of others. When we are lost, most of us look to see what others are doing. Even then, we often get lost. Sometimes even whole groups of friends and colleagues get lost together. Consequently, the hardest thing about science is what it demands of us in terms of our ability to make the right choice in the face of incomplete information. This requires characteristics not easily measured by tests, such as intuition and a person's faith in themself. Einstein knew this, which is why he told John Wheeler, in a remark that Wheeler has often repeated, how much he admired Newton's courage and judgement in sticking with the idea of absolute space and time even though all his colleagues told him it was absurd. The idea *is* absurd, as Einstein knew better than anyone. But absolute space and time was what was required to make progress at the time, and to see this was perhaps Newton's greatest achievement.

Einstein himself is often presented as the prime example of someone who did great things alone, without the need for a community. This myth was fostered, perhaps even deliberately, by those who have conspired to shape our memory of him. Many of us were told a story of a man who invented general relativity out of his own head, as an act of pure individual creation, serene in his contemplation of the absolute as the First World War raged around him.

It is a wonderful story, and it has inspired generations of us to wander with unkempt hair and no socks around shrines like Princeton and Cambridge, imagining that if we focus our thoughts on the right question we could be next great scientific icon. But this is far from what happened. Recently my partner and I were lucky enough to be shown pages from the actual notebook in which Einstein invented general relativity, while it was being prepared for publication by a group of historians working in Berlin. As working physicists it was clear to us right away what was happening: the man was confused and lost – very lost. But he was also a very good physicist (though not, of course, in the sense of the mythical saint who could perceive truth directly). In that notebook we could see a very good physicist exercising the same skills and strategies, the mastery of which made Richard Feynman such a great physicist. Einstein knew what to do when he was lost: open his notebook and attempt some calculation that might shed some light on the problem.

So we turned the pages with anticipation. But still he gets nowhere. What does a good physicist do then? He talks with his friends. All of a sudden a name is scrawled on the page: 'Grossmann!!!' It seems that his friend has told Einstein about something called the curvature tensor. This is the mathematical structure that Einstein had been seeking, and is now understood to be the key to relativity theory.

Actually I was rather pleased to see that Einstein had not been able to invent the curvature tensor on his own. Some of the books from which I had learned relativity had seemed to imply that any competent student should be able to derive the curvature tensor given the principles Einstein was working with. At the time I had had my doubts, and it was reassuring to see that the only person who had ever actually faced the problem without being able to look up the answer had not been able to solve it. Einstein had to ask a friend who knew the right mathematics.

The textbooks go on to say that once one understands the curvature tensor, one is very close to Einstein's theory of gravity. The questions Einstein is asking should lead him to invent the theory in half a page. There are only two steps to take, and one can see from this notebook that Einstein has all the ingredients. But could he do it? Apparently not. He starts out promisingly, then he makes a mistake. To explain why his mistake is not a mistake he invents a very clever argument. With falling hearts, we, reading his notebook, recognize his argument as one that was held up to us as an example of how *not* to think about the problem. As good students of the subject we know that the argument being used by Einstein is not only wrong but absurd, but no one told us it was Einstein himself who invented it. By the end of the notebook he has convinced himself of the truth of a theory that we, with more experience of this kind of stuff than he or anyone could have had at the time, can see is not even mathematically consistent. Still, he convinced himself and several others of its promise, and for the next two years they pursued this wrong theory. Actually the right equation was written down, almost accidentally, on one page of the notebook we looked at it. But Einstein failed to recognize it for what it was, and only after following a false trail for two years did he find his way back to it. When he did, it was questions his good friends asked him that finally made him see where he had gone wrong.

Nothing in this notebook leads us to doubt Einstein's greatness – quite the contrary, for in this notebook we can see the trail followed by a great human being whose courage and judgement are strong enough to pull him through a thicket of confusion from which few others could have emerged. Rather, the lesson is that trying to invent new laws of physics is hard. Really hard. No one knew better than Einstein that it requires not only intelligence and hard work, but equal helpings of insight, stubbornness, patience and character. This is why all scientists work in communities. And that makes the history of science a human story. There can be no triumph without an equal amount of foolishness. When the problem is as hard as the invention of quantum gravity, we must respect the efforts of others even when we disagree with them. Whether we travel in small groups of friends or in large convoys of hundreds of experts, we are all equally prone to error.

Another moral has to do with why Einstein made so many mistakes in his struggle to invent general relativity. The lesson he had such trouble learning was that space and time have no absolute meaning and are nothing but systems of relations. How Einstein himself learned this lesson, and by doing so invented a theory which more than any other realizes the idea that space and time are relational, is a beautiful story. But it is not my place here to tell it – that must be left to historians who will tell it right.

The subject of this chapter is string theory, and I begin it with these reflections for two reasons. First, because the main thing that is wrong with string theory, as presently formulated, is that it does not respect the fundamental lesson of general relativity that spacetime is nothing but an evolving system of relationships. Using the terminology I introduced in earlier chapters, string theory is background dependent, while general relativity is background independent. At the same time, string theory is unlikely to be in its final form. Even if, as is quite possible, string theory is ultimately reformulated in a background independent form, history may record that Einstein's view of Newton applies also to the string theorists: when it was necessary to ignore fundamental principle in order to make progress, they had the courage and the judgement to do so.

The story of string theory is not easy to tell, because even now we do not really know what string theory is. We know a great deal about it, enough to know that it is something really marvellous. We know much about how to carry out certain kinds of calculations in string theory. Those calculations suggest that, at the very least, string theory may be part of the ultimate quantum theory of gravity. But we do not have a good definition of it, nor do we know what its fundamental principles are. (It used to be said that string theory was part of twenty-first-century mathematics that had fallen by luck into our hands in the twentieth century. This does not sound quite as good now as it used to.) The problem is that we do not yet have string theory expressed in any form that could be that of a fundamental theory. What we have on paper cannot be considered to be the theory itself. What we have is no more than a long list of examples of solutions to the theory; what we do not yet have is the theory they are solutions of. It is as if we had a long list of solutions to the Einstein equations, without knowing the basic principles of general relativity or having any way to write down the actual equation that defines the theory.

Or, to take a simpler example, string theory in its present form most likely has the same relationship to its ultimate form as Kepler's astronomy had to Newton's physics. Johannes Kepler discovered that the planets travel along elliptical orbits, and he was able to use this principle together with two other rules he discovered to write down an infinite number of possible orbits. But it took Newton to discover the reason why the planetary orbits are ellipses. This allowed him to unify the explanation of the motions of the planets with many other observed motions, such as the parabolic trajectories that Galileo had discovered are followed by projectiles on the Earth. Many more examples of solutions to string theory have recently been discovered, and the virtuosity required to construct these solutions in the absence of a fundamental principle is truly humbling. This has made it possible to learn a lot about the theory, but so far, at least, it does not suffice to tell us what the theory is. No one has yet had that vital insight that will make it possible to jump from the list of solutions to the principles of the theory.

Let us begin, then, with what we do know about string theory, for these are reasons enough to take it seriously. Quantum theory says that for every wave there is an associated particle. For electromagnetic waves there is the photon. For electrons there is the electron wave (the wave-function). The wave doesn't even have to be something fundamental. When I strike a tuning fork I set up waves that travel up and down it: these are sound waves travelling in metal. Quantum theory associates a particle with such sound waves; it is called a phonon. Suppose I disturb the empty space around us by making a gravitational wave. This can be done by waving around anything with mass – one of my arms will do, or a pair of neutron stars. A gravitational wave can be understood as a tiny ripple moving against a background, which is the empty space.

The particle associated with gravitational waves is called the graviton. No one has ever observed a graviton. It is hard enough even to detect a gravitational wave, as they interact only very weakly with matter. But as long as quantum theory applies to gravitational waves, gravitons must exist. We know that gravitons must interact with matter, for when anything massive oscillates it produces gravitational waves. Quantum theory says that, just as there are photons associated with light, there must be gravitons associated with gravitational waves.

We know that two gravitons will interact with each other. This is because gravitons interact with anything that has energy, and gravitons themselves carry energy. As with the photon, the energy of a graviton is proportional to its frequency, so the higher the frequency of a graviton, the more strongly it will interact with another graviton. We can then ask what happens when two gravitons interact. We know that they will scatter from each other, changing their trajectories. A good quantum theory of gravity must be able to predict what will happen whenever two gravitons interact. It ought to be able to produce an answer no matter how strong the waves are and no matter what their frequencies are. This is just the kind of question that we know how to approach in quantum theory. For example, we know that photons will interact with any charged particle, such as an electron. We have a good theory of the interactions of photons and electrons, called *quantum electrodynamics*, QED for short. It was developed by Richard Feynman, Julian Schwinger, Sinitiro Tomonaga and others in the late 1940s. QED makes predictions about the scattering of photons and electrons and other charged particles that agree with experiment to an accuracy of eleven decimal places.

Physics, like the other sciences, is the art of the possible. So I must add a rider here, which is that we do not really understand QED. We know the principles of the theory and we can deduce from them the basic equations that define the theory. But we cannot actually solve these equations, or even prove that they are mathematically consistent. Instead, to make sense of them we have to resort to a kind of subterfuge. We make some assumptions about the nature of the solutions – which, after more than fifty years, are still unproved – and these lead us to a procedure for calculating approximately what happens when photons and electrons interact. This procedure is called *perturbation theory*. It is very useful in that it does lead to answers that agree very precisely with experiment. But we do not actually know whether the procedure is consistent or not, or whether it accurately reflects what a real solution to the theory would predict. String theory is presently understood mainly in the language of this approximation procedure. It was invented by modifying the approximation procedure, rather than the theory. This is how people were able to invent a theory which is understood only as a list of solutions.

Perturbation theory is actually quite easy to describe. Thanks to Feynman, there is a simple diagrammatic means for understanding it. Picture a world of processes in which three things can happen. An electron may move from point A at one time to point B at another. We can draw this as a line, as in Figure 33. A photon may also travel, which is indicated by a dotted line in the figure. The only other thing that may happen is that an electron and a photon interact, which is indicated by the point where a photon line meets an electron line. To compute what happens when two electrons meet, one simply draws all the things that can happen, beginning with two electrons entering the scene, and ending with two electrons leaving. There are an infinite number of such processes, and we see a few of them in Figure 34. Feynman taught us to associate with each diagram the probability (actually the quantum amplitude, whose square is the probability) of that process. One can then work out all the predictions of the theory.

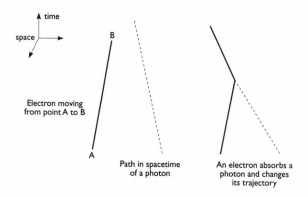

time

space

B

Electron moving
from point A to B

A

Path in spacetime
of a photon

An electron absorbs a
photon and changes
its trajectory

FIGURE 33

The basic processes in the theory of electrons and photons (called quantum electrodynamics, or QED for short.) Electrons and photons can move freely in spacetime, or they can interact in events in which an electron absorbs or emits a photon.

In the language of these diagrams, now known appropriately as *Feynman diagrams*, it is very easy to explain what string theory is. The basic postulate of the theory is that there are no particles, only strings moving in space. A string is just a loop drawn in space. It is not made of anything, just as a particle is thought of as a point and nothing else. There is only one kind of string, and the different kinds of particle are postulated to be nothing but different modes of vibration of these loops. So, as shown in Figure 35, photons and electrons are to be thought of just as different ways in which a string can vibrate. When a string moves in time it makes a tube rather than a line (Figure 35). Two strings can also join and merge into one (Figure 36), or one string can split into two. All the interactions that occur in nature, including those of photons and electrons, can be interpreted in terms of these splittings and joinings. We can see from these pictures that string theory gives a very satisfactory unification and simplification of the physical processes represented in Feynman diagrams. Its main virtue is that it gives a simple way of finding theories that make consistent physical predictions.

The trouble with Feynman's method is that it always leads

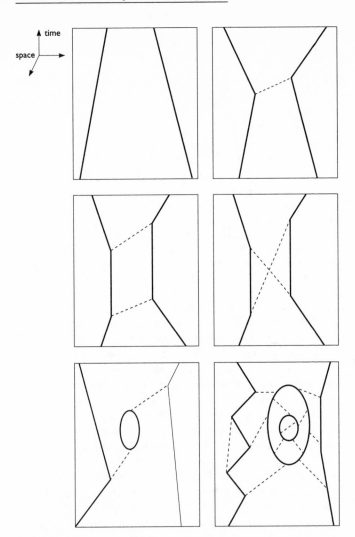

FIGURE 34

The processes illustrated in Figure 33 are put together to make Feynman diagrams, which are pictures of the possible ways a process can happen. Shown here are some of the ways in which two electrons can interact simply by absorbing and emitting photons. Each one is a story that is a possible piece of the history of a universe.

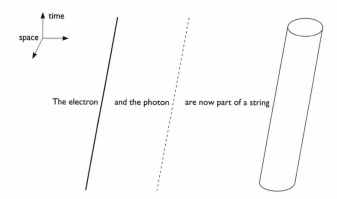

FIGURE 35

In string theory there is only one kind of thing that moves, and that is a string – a loop drawn in space. Different modes of vibration of the string behave like the different kinds of elementary particle.

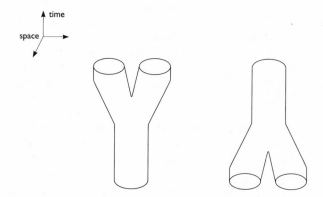

FIGURE 36

All the different kinds of interaction between particles are interpreted in string theory in terms of the splitting and joining of strings.

to infinite expressions. This is because there are loops in the diagrams where particles are created, interact, and are then destroyed. These are called *virtual particles* because they exist only for a very short time. According to the uncertainty principle, because virtual particles live for a very short time they can have any energy, as the conservation of energy is suspended during their brief lives. This creates big problems. One has to add up all the diagrams to get the overall probability for the process to occur, but if some particles can have any energy between zero and infinity, then the list of possible processes one has to add up will be infinite. This leads to mathematical expressions that are no more than complicated ways of writing the number infinity. As a result, Feynman's method seems at first to give nonsensical answers to questions about the interactions of electrons and photons.

Quite ingeniously, Feynman and others discovered that the theory was giving silly answers to only a few questions, such as 'What is the mass of the electron?' and 'What is its charge?' The theory predicts that these are infinite! Feynman figured out that if one simply crosses out these infinite answers wherever they appear, and substitutes the right, finite answer, the answers to all other questions become sensible. All the infinite expressions can be removed if one forces the theory to give the right answer for the mass and the charge of the electron. This procedure is called *renormalization*. When it works for a theory, that theory is called renormalizable. The procedure works very well for quantum electrodynamics. It also works for quantum chromodynamics, and for the Weinberg–Salam theory, which is our theory of radioactive decay. When this procedure does not work, we say that a theory is not renormalizable – the method fails to give a sensible theory. This is actually the case for most theories; only certain special ones can be made sense of by these methods.

The most important theory that cannot be made sense of in this way is Einstein's theory of gravity. The reason has to do with the fact that arbitrarily large energies can appear in the particles moving inside the diagrams. But the strength of the gravitational force is proportional to the energy, because energy is mass, from Einstein, and gravity pulls on mass, from Newton. So the diagrams with larger energies give correspondingly larger effects. But according to the theory, the energies inside the diagrams can be arbitrarily large. The result is a kind of runaway feedback process in which we lose all control over what is happening inside the diagrams. No one has ever found a way to describe a gravitational theory in the language of particles moving around Feynman diagrams. But in string theory one can make sense of the effects of gravity. This is one of its great achievements. As with the older theories, there are many string theory variants that lead to infinite expressions for every physical process, and these must be discarded. What is left is a set of theories that have no infinities at all. One does not have to play any games to isolate infinite expressions for masses and throw them out. There are just two possible kinds of string theory: inconsistent and consistent. And all the consistent ones appear to give finite and sensible expressions for all physical quantities.

The list of consistent string theories is very long. There are consistent string theories in all dimensions from one to nine. In nine dimensions there are five different kinds of consistent string theory. When we get down to the three-dimensional world we seem to live in, there are at least hundreds of thousands of different consistent string theories. Most of these theories come with free parameters, so they do not make unique predictions for things like the masses of the elementary particles. Each consistent string theory is very tightly structured. Because all the different kinds of particle arise from vibrations of the same fundamental objects, one is not generally free to choose which particles are described by the theory. There are an infinite number of possible vibrations and hence of possible particles, although most of them will have energies which are too large to observe. Only the lowest modes of vibration correspond to particles with masses we could observe. A remarkable fact is that the particles that correspond to the lowest modes of vibration of a string always include the broad categories of particles and forces we do observe. The other modes of vibration correspond to particles with masses of around 10^{19} times the mass of the proton. This is the Planck mass, which is the mass of a black hole the size of a Planck length.

However, there still are issues which must be addressed if string theory is to describe our universe. Many string theories predict the existence of particles which have so far not been seen. Many have problems keeping the strength of the gravitational force from varying in space and time. And almost all consistent string theories predict symmetries among their particles beyond those that are seen. The most important of these are *supersymmetries*.

Supersymmetry is an important idea, so it is worth while making a detour here to discuss it. To understand super-symmetry one must know that elementary particles come into two general types: bosons and fermions. Bosons, which include photons and gravitons, are particles whose angular momentum, when measured in units of Planck's constant, are simple integers. Fermions, which include electrons, quarks and neutrinos, have angular momenta that come in units of one-half. Fermions also satisfy the Pauli exclusion principle, which states that no two of them can be put in the same state. Supersymmetry requires fermions and bosons to come in pairs consisting of one of each, with the same mass. This is definitely not observed in nature. If there were such things as bosonic electrons and quarks, the world would be a very different place, for the Pauli exclusion principle would have no force, and no form of matter would be stable. If super-symmetry is true of our world, then it has been *spontaneously broken*, which is to say that the background fields must confer a large mass on one member of each pair and not on the other. The only reason to entertain the idea of such a strange symmetry is that it seems to be required for most, if not all, versions of string theory to give consistent answers.

The search for evidence of supersymmetry is a major priority of experiments now under way at particle accelera-tors. String theorists very much hope that evidence for supersymmetry will be found. If supersymmetry is not found experimentally, it would still be possible to concoct a string theory that agrees with experiment, but this would be a less happy outcome than if experimental support for supersym-metry were forthcoming.

There is obviously something very wonderful about string theory. Among its strong points are the natural way it leads to a unification of all particles and forces, and the fact that there are many consistent string theories that include gravity. String theory is also the perfect realization of the hypothesis of duality discussed in Chapter 9. Also, it cannot be over-emphasized that in the language in which it is understood – that of diagrams corresponding to quantum particles moving against a background spacetime – string theory is the only known way of consistently unifying gravity with quantum theory and the other forces of nature.

What is very frustrating is that in spite of this, string theory does not seem to fully incorporate the basic lesson of general relativity, which is that space and time are dynamical rather than fixed, and relational rather than absolute. In string theory, as it has so far been formulated, the strings move against a background spacetime which is absolute and fixed. The geometry of space and time is usually presumed to be fixed for ever; all that happens is that some strings move against this fixed background and interact with one another. But this is wrong, because it replicates the basic mistake of Newtonian physics in treating space and time as a fixed and unchanging background against which things move and interact. As I have already emphasized, the right thing to do is to treat the whole system of relationships that make up space and time as a single dynamical entity, without fixing any of it. This is how general relativity and loop quantum gravity work.

Still, science is not made from absolutes. The progress of science is based on what is possible, which means that it often makes sense to do what is practical, even if it seems to go against established principles. For this reason, even if it is ultimately wrong, it may still be useful to follow the background dependent approach as far as it will go, to see whether there is a consistent picture in which we can answer questions such as what happens when two gravitons moving in empty spacetime scatter from each other. As long as we remember that such a picture can give at best an approximate description this can be an important and necessary step in the discovery of the quantum theory of gravity.

Another main shortcoming of string theory is that is not one theory, but a whole class of theories, so it does not lead to many predictions about the elementary particles. This short-coming is closely related to the problem of background dependence. Each string theory moves against a different background spacetime, so to define a string theory one must first fix the dimension of space and the geometry of spacetime. In many cases space has more dimensions than the three we observe. This is explained by the hypothesis that in our universe the extra dimensions are curled up too tightly for us to perceive directly. We say that the extra six dimensions have been *compactified*. Since string theory is simplest if the world has nine spatial dimensions, this leads to a picture in which many of the different consistent string theories in three dimensions can be understood as arising from different ways of choosing the structure of a hidden six-dimensional space.

There are at least hundreds of thousands of ways in which the six extra dimensions may be compactified. Each way corresponds to a different geometry and topology for the extra six dimensions. As a result there are at least that many different string theories that are consistent with the basic observation that the world has three large spatial dimensions. Furthermore, each of these theories has a set of parameters that describe the size and other geometric properties of the six compactified dimensions. These turn out to influence the physics that we see in the three-dimensional world. For example, the geometry of the extra dimensions influences the masses and the strengths of the interactions of the elementary particles we observe.

It is most likely irrelevant whether these extra dimensions exist in any literal sense. If one is drawn to a picture of our three-dimensional 'reality' embedded in some higher-dimensional realm, then one can believe in the extra dimensions, at least as long as one is working in this background dependent picture. But these extra dimensions can also be seen as purely theoretical devices which are useful for understanding the list of consistent string theories in three dimensions. As long as we stay on the background dependent level, it does not really matter.

As a result, although it is a unified theory, string theory in its present form makes few predictions about the physics we actually observe. Many different scenarios for what the new, more powerful particle accelerators will find are consistent with one version of string theory or another. Thus, not only does string theory lack experimental confirmation, but it is hard to imagine an experiment that could be done in the next several decades that could definitively confirm or reject it. Nor is there anything special, from the point of view of string theory, about having six out of nine dimensions compactified while the other three are left large. String theory can easily describe a world in which any number of dimensions, from nine down to none at all, are left large.

String theory thus indicates that the world we see provides only a sparse and narrow sampling of all possible physical phenomena, for if true it tells us that most of the dimensions and most of the symmetry of the world are hidden. Still, many people do believe in it. This is partly because, however incomplete its present formulation may be, string theory remains the one approach that unifies gravity with the other forces consistently at a background dependent level.

The main problem in string theory, then, is how to see beyond it to a theory which will incorporate the successes of string theory while avoiding its weaknesses. One approach to this problem begins with the following question. What if there were a single theory that unified the different string theories by interpreting each of its solutions as one of the consistent string theories? The different string theories, together with the spacetimes they live in, will not be put in as absolutes. Rather they will all arise from solutions of this new theory. Note that the new theory could not be formulated in terms of any objects moving against a fixed spacetime background, because its solutions would include all the possible background spacetimes. The different solutions of this fundamental theory would be analogous to the different spacetimes which are all solutions to the equations of general relativity.

Now we can argue by analogy in the following way. Let us take any spacetime which is a solution to the Einstein equations, and wiggle some matter within it. This will generate gravitational waves. These waves move on the original spacetime like ripples moving on the surface of a pond. We can make ripples in the solution of our fundamental theory in the same way. What if these gave rise not to waves moving on the background, but to strings? This may be hard to visualize, but remember that according to the hypothesis of duality strings are just a different way of looking at a field, like the electric field. And if we wiggle a field we get waves. The wiggles in the electric and magnetic field are after all nothing but light. But if duality is true, there must be a way to understand this in terms of the motion of strings through space.

If this picture is correct, then each string theory is not really a theory in its own right. It is no more than an approximate description of how ripples may move against a background spacetime which itself is a solution to another theory. That theory would be some extension of general relativity, formulated in terms that were relational and background independent.

This hypothesis would, if true, explain why there are so many different string theories. The solutions to the fundamental theory will define a large number of different possible universes, each described in terms of a different space and time.

It remains only to construct this single, unifying string theory. This is a project that a few people are working hard on, and I must confess it is something I also am spending a lot of time on. There is presently no agreed upon form of this theory, but at least we have a name for it – we call it M theory. No one knows what the M stands for, which we feel is appropriate for a theory whose existence has so far only been conjectured.

These days, string theorists spend much of their time looking for evidence that M theory exists. One strategy which has been very successful is to look for relationships between different string theories. A number of cases have been found in which two apparently different versions of string theory turn out to describe exactly the same physical phenomena. (In some cases this is seen directly; in others the coincidence is apparent only certain approximations or from studying simplified versions of the theories in which extra symmetries have been imposed.) These relationships suggest that the different string theories are part of a larger theory. The information about these relationships can be used to learn something about the structure M theory must have, if it exists. For example, it gives us some information about the symmetries that M theory will have. These are symmetries that extend the idea of duality in a major way, which could not be done within any single string theory.

Another very important question is whether M theory describes a universe in which space and time are continuous or discrete. At first it seems that string theory points to a continuous world, because it is based on a picture of strings moving continuously through space and time. But this turns out to be misleading, for when looked at closely string theory seems to be describing a world in which space has a discrete structure.

One way to see the discreteness is to study strings on a space that has been wrapped up, so that one dimension forms a circle (Figure 37). The circle which has been wrapped up has radius R. You might think that the theory would get into trouble if we allowed R to get smaller and smaller. But string theory turns out to have the amazing property that what

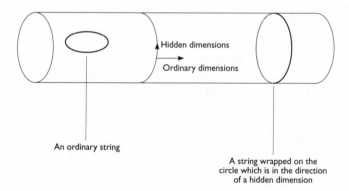

FIGURE 37

A cylinder is a two-dimensional space in which one direction is a circle. We see a string wrapped on the circle. This is typical of ideas of how the extra dimensions are hidden; the horizontal direction is typical of the three ordinary directions, while the vertical direction stands for one of the hidden dimensions. Time is not indicated here.

happens when R becomes very small is indistinguishable from what happens when R becomes very large. The result is that there is a smallest possible value for R. If string theory is right, then the universe cannot be smaller than this.

There is a pretty simple explanation for this, which I hope will at least give you a taste of the kind of reasoning that permeates the study of string theory. The reason why R has a smallest possible value has to do with the fact that there are two different things a string can do when wrapped around a cylinder (it is said to have two degrees of freedom). First, it can vibrate, like a guitar string. Since the radius of the cylinder is fixed there will be a discrete series of modes in which the string can vibrate. But the string has another degree of freedom, because one can vary the number of times it is wrapped around the cylinder. Thus there are two numbers that characterize a string wrapped around a cylinder: the mode number and the number of times it is wrapped.

It turns out that if one tries to decrease the radius of the cylinder, R, below a certain critical value, these two numbers just trade places. A string in the 3rd mode of vibration wrapped 5 times around a cylinder with R slightly smaller than the critical value becomes indistinguishable from a string wrapped 3 times around a cylinder slightly larger than the critical value, when it is in the 5th mode of vibration. The effect is that every mode of vibration of a string on a small cylinder is indistinguishable from a different mode of a string wrapped on a large cylinder. Since we cannot tell them apart, the modes of strings wrapped around small cylinders are redundant. All the states of the theory can be described in terms of cylinders larger than the critical value.

Another way to see the discreteness is to imagine a string going by at very nearly the speed of light. It would appear to contain a set of discrete elements, each of which carries a certain fixed amount of momentum. These are called *string bits*, and they are shown in Figure 38. The more momentum a string has, the longer it is, so there is a limit to the size of an object that can be resolved by looking at it with a string. But since, according to string theory, all the particles in nature are actually made up of strings, then, if the theory is right, there is a smallest size. Just as there is a smallest piece of silver, which is a silver atom, there is a smallest possible process that can propagate, and that is a string bit.

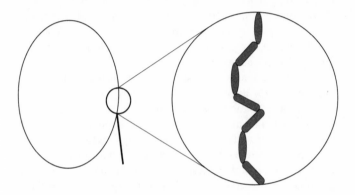

FIGURE 38
A string seen through a Planck-scale magnifying glass is found to consist of discrete bits, rather like a wooden toy snake.

There turns out to be a simple way to express the fact that there is a minimum size that can be probed in string theory. In ordinary quantum theory the limitations to what can be observed are expressed in terms of the Heisenberg uncertainty principle. This says that

$$\Delta x > (h/\Delta p)$$

where Δx is the uncertainty in position, h is Planck's famous constant and Δp is the uncertainty in momentum. String theory amends this equation to

$$\Delta x > (h/\Delta p) + C\Delta p$$

where C is another constant that has to do with the Planck scale. Now, without this new term one can make the uncertainty in position as small as one likes, by making the uncertainty in momentum large. With the new term in the equation one cannot do this, for when the uncertainty in momentum becomes large enough the second term comes in and forces the uncertainty in position to start to increase rather than decrease. The result is that there is a minimum value to the uncertainty in position, and this means that there is an absolute limit to the precision with which any object can be located in space.

This tells us that M theory, if it exists, cannot describe a world in which space is continuous and one can pack an infinite amount of information into any volume, no matter how small. This suggests that whatever it is, M theory will not be some direct extension of string theory, as it will have to be formulated in a different conceptual language. The present formulation of string theory is likely, then, to be a transitional stage in which elements of a new physics are mixed up with the old Newtonian framework, according to which space and time are continuous, infinitely divisible and absolute. The problem that remains is to separate out the old from new and find a coherent way to formulate a theory using only those principles that are supported by the experimental physics of the twentieth and twenty-first centuries.

III
THE PRESENT
FRONTIERS

....

CHAPTER 12

THE HOLOGRAPHIC PRINCIPLE

In Part II we looked at three different approaches to quantum gravity: black hole thermodynamics, loop quantum gravity and string theory. While each takes a different starting point, they all agree that when viewed on the Planck scale, space and time cannot be continuous. For seemingly different reasons, at the end of each of these roads one reaches the conclusion that the old picture according to which space and time are continuous must be abandoned. On the Planck scale, space appears to be composed of fundamental discrete units.

Loop quantum gravity gives us a detailed picture of these units, in terms of spin networks. It tells us that areas and volumes are quantized and come only in discrete units. String theory at first appears to describe a continuous string moving in a continuous space. But a closer look reveals that a string is actually made of discrete pieces, called string bits, each of which carries a discrete amount of momentum and energy. This is expressed in a simple and beautiful way as an extension of the uncertainty principle, which tells us that there is a smallest possible length.

Black hole thermodynamics leads to an even more extreme conclusion, the Bekenstein bound. According to this principle the amount of information that can be contained in any region is not only finite, it is proportional to the area of the boundary of the region, measured in Planck units. This implies that the world must be discrete on the Planck scale, for were it continuous any region could contain an infinite amount of information.

It is remarkable that all three roads lead to the general conclusion that space becomes discrete on the Planck scale. However, the three different pictures of quantum spacetime that emerge seem rather different. So it remains to join these pictures together to make a single picture which, when we understand it, will become the one final road to quantum gravity.

At first it may not be obvious how to do this. The three different approaches investigate different aspects of the world. Even if there is one ultimate theory of quantum gravity, there will be different physical regimes, in which the basic principles may manifest themselves differently. This seems to be what is happening here. The different versions of discreteness arise from asking different questions. We would find an actual contradiction only if, when we asked the same question in two different theories, we got two different answers. So far this has not happened, because the different approaches ask different kinds of question. It is possible that the different approaches represent different windows onto the same quantum world – and if this is so, there must be a way of unifying them all into a single theory.

If the different approaches are to be unified, there must be a principle which expresses the discreteness of quantum geometry in a way that is consistent with all three approaches If such a principle can be found, then it will serve as a guide to combining them into one theory. In fact, just such a principle has been proposed in recent years. It is called the *holographic principle.*

Several different versions of this principle have been proposed by different people. After a lot of discussion over the last few years there is still no agreement about exactly what the holographic principle means, but there is a strong feeling among those of us in the field that some version of the holographic principle is true. And if it is true, it will be the first principle which makes sense only in the context of a quantum theory of gravity. This means that even if it is presently understood as a consequence of the principles of general relativity and quantum theory, there is a chance that in the end the situation will be reversed and the holographic principle will become part of the foundations of physics, from which quantum theory and relativity may both be deduced as special cases.

The holographic principle was inspired first of all by the Bekenstein bound, which we discussed in Chapter 8. Here is one way to describe the Bekenstein bound. Consider any physical system, made of anything at all – let us call it The Thing. We require only that The Thing can be enclosed within a finite boundary, which we shall call The Screen (Figure 39). We would like to know as much as possible about The Thing. But we cannot touch it directly – we are restricted to making measurements of it on The Screen. We may send any kind of radiation we like through The Screen, and record whatever changes result on The Screen. The Bekenstein bound says that there is a general limit to how many yes/no questions we can answer about The Thing by making observations through

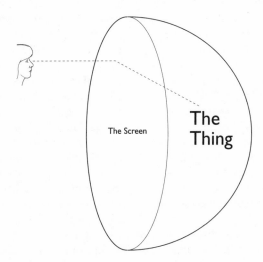

FIGURE 39
The argument for the Bekenstein bound. We observe The Thing through The Screen, which limits the amount of information we can receive about The Thing to what can be represented on The Screen.

The Screen that surrounds it. The number must be less than one-quarter of the area of The Screen, in Planck units. What if we ask more questions? The principle tells us that either of two things must happen. Either the area of the screen will increase, as a result of doing an experiment that asks questions beyond the limit; or the experiments we do that go beyond the limit will erase, or invalidate, the answers to some of the previous questions. At no time can we know more about The Thing than the limit, imposed by the area of The Screen.

What is most surprising about this is not just that there is a limit on the amount of information that can be coded into The Thing – after all, if we believe that the world has a discrete structure then this is exactly what we should expect. It is just that we would normally expect the amount of information that can be coded into The Thing to be proportional to its volume, not to the area of a surface that contains it. For example, suppose that The Thing is a computer memory. If we continue to miniaturize computers more and more, we shall eventually be building them purely out of the quantum geometry in space – and that has to be the limit of what can be done. Imagine that we can then build a computer memory out of nothing but the spin network states that describe the quantum geometry of space. The number of different such spin network states can be shown to be proportional to volume of the world that state describes (The reason is that there are so many states per node, and the volume is proportional to the number of nodes.) The Bekenstein bound does not dispute this, but it asserts that the amount of information that we outside observers could extract is proportional to the area and not the volume. And the area is proportional not to the number of nodes of the network, but to the number of edges that go through the screen (Figure 40). This tells us that the most efficient memory we could construct out of the quantum geometry of space is achieved by constructing a surface and putting one bit of memory in every region 2 Planck lengths on a side. Once we have done this, building the memory into the third dimension will not help.

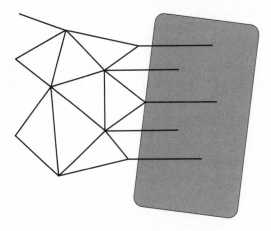

FIGURE 40

A spin network, which describes the quantum geometry of space, intersects a boundary such as a horizon in a finite number of points. Each intersection adds to the total area of the boundary.

This idea is very surprising. If it is to be taken seriously, there had better be a good reason for it. In fact there is, for the Bekenstein bound is a consequence of the second law of thermodynamics. The argument that leads from the laws of thermodynamics to the Bekenstein bound is not actually very complicated. Because of its importance I give a form of it in the box on the next page.

There are at least two more good reasons to believe in the Bekenstein bound. One is that the relationship between Einstein's theory and the bound can be turned around. In the argument for the Bekenstein bound as I present it in the box, the bound is partly a consequence of the equations of Einstein's general theory of relativity. But, as Ted Jacobson has shown in a justly famous paper, the argument can be turned on its head so that the equations of Einstein's theory can be derived by assuming that the laws of thermodynamics and the Bekenstein bound are true. He does this by showing that the area of The Screen must change when energy flows through it, because the laws of thermodynamics require that some entropy flows along with the energy. The result is that

Argument for the Bekenstein bound

Let us suppose that The Thing is big enough to be described both in terms of an exact quantum description and in terms of an averaged, macroscopic description. We shall argue by contradiction, which means that we first assume the opposite of what we are trying to show. Thus we assume that the amount of information required to describe The Thing is much larger than the area of The Screen. For simplicity, we assume that The Screen is spherical.

We know that The Thing is not a black hole, because we know that the entropy of any black hole that can fit into The Screen must be equivalent to an area less than that of the screen. But in this case its entropy must be lower than the area of the screen, in Planck units. If we assume that the entropy of a black hole counts the number of its possible quantum states, this is much less than the information contained in The Thing.

It then follows (from a theorem of classical general relativity) that The Thing has less energy than a black hole that would just fit inside The Screen. Now, we can slowly add energy to The Thing by dripping it slowly through the screen. We shall reach some point by which we shall have given it so much energy that, by that same theorem, it must collapse to a black hole. But then we know that its entropy is given by one-quarter of the area of the screen. Since that is lower than the entropy of The Thing initially, we have managed to lower the entropy of a system. This contradicts the second law of thermodynamics.

We dripped the energy in slowly to ensure that nothing surprising happens outside The Screen which might increase the entropy strongly somewhere else. There seem to be no loopholes in this argument. Therefore, if we believe the second law of thermodynamics, we must believe that the most entropy that we, outside the Screen, can attribute to The Thing is one-quarter of the area of The Screen. And because entropy is a count of answers to yes/no questions, this implies the Bekenstein bound as we have stated it.

the geometry of space, which determines the area of The Screen, must change in response to the flow of energy. Jacobson shows that this actually implies the equations of Einstein's theory.

Another reason to believe the Bekenstein bound is that it can be derived directly from loop quantum gravity. To do this one only has to study the problem of how a screen is described by the quantum theory. As shown in Figure 40, in loop quantum gravity a screen will be pierced by edges of a spin network. Each edge that intersects the screen contributes to the total area of the screen. It turns out that each edge that is added also increases the amount of information that can be stored in a quantum theoretic description of the screen. We can add more edges, but the information a screen can store cannot increase faster than its area. This is just what is required by the Bekenstein bound.

Perhaps the first person to realize the radical implications of the Bekenstein bound was Louis Crane. He deduced from it that quantum cosmology must be a theory of the information exchanged between subsystems of the universe, rather than a theory of how the universe would look to an outside observer. This was the first step towards the relational theories of quantum cosmologies later developed by Carlo Rovelli, Fotini Markopoulou and myself. Gerard 't Hooft later began to think about the horizon of a black hole as something like a computer, along the lines I have described. He proposed the first version of the holographic principle and gave it its name. It was then quickly championed by Leonard Susskind, who showed how it could be applied to string theory. Since then at least two other versions of the holographic principle have been proposed. So far there is no consensus on which is right. I shall explain two of the versions, which are called the strong and weak holographic principles.

The idea of the *strong holographic principle* is very simple. Since the observer is restricted to examining The Thing by making observations through The Screen, all of what is observed could be accounted for if one imagined that, instead of The Thing, there was some physical system defined on the screen itself (Figure 41). This system would be described by a

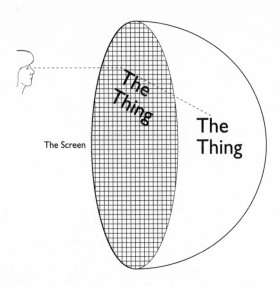

FIGURE 41

The Screen is like a television set with pixels measuring 2 Planck lengths on each side. One can only see as much information about the world beyond The Screen as can be represented on it.

theory which involved only The Screen. This 'screen theory' might describe The Screen as something like a quantum computer, with one bit of memory for every pixel, each pixel being 2 Planck lengths on each side. Now suppose that the observer sends some signal through The Screen, which interacts with The Thing. The result is a signal which comes back out through The Screen. As far as the observer is concerned, the same thing would happen if the light interacted with the quantum computer on The Screen and returned a suitable signal. The point is that there is no way for the observer to tell if they were interacting with The Thing itself, or merely with its image, represented as a state of the screen theory. If the screen theory were suitably chosen, or the computer representing the information on the screen suitably programmed, the laws of physics holding inside the screen could equally well be represented by the response of The Screen to the observer.

In this form, the holographic principle states that the most succinct description that can be given of the part of the world that lies on the other side of any surface is actually a description of how its image evolves on that surface. This might seem weird, but the important thing is the way it relies on the Bekenstein bound. The Screen description is adequate because no more information can be gained about The Thing than can ever be represented by the state of the pixels on The Screen. The strong form of the holographic principle says that the world is such that the physical description of any object in nature can equally well be represented by the state of such a computer, imagined to exist on a surface surrounding it. That is, for every set of true laws that might hold inside The Screen, there is a way to program the computer representing the screen theory so that it reproduces all true predictions of those laws.

This is weird enough, but it does not go as far as it might. The problem is that it describes the world in terms of things. But remember, in Chapter 4 I argued that when we get down to the fundamental theory there will be no things, only processes. If we believe this, we cannot believe in any principle which expresses the world in terms of things. We should reformulate the principle so that it makes references only to processes. This is what the *weak holographic principle* does. It states that we are mistaken to think that the world consists of Things that occupy regions of space. Instead, all that there exists in the world are Screens, on which the world is represented. That is, it does not posit that there are two things, bulky things, and images or representations of them on their surfaces. It posits that there is only one kind of thing – representations by which one set of events in the history of the universe receives information about other parts of the world.

In such a world, nothing exists except processes by which information is conveyed from one part of the world to another. And the area of a screen – indeed, the area of any surface in space – is really nothing but the capacity of that surface as a channel for information. So, according to the weak holographic principle, space is nothing but a way of talking about all the different channels of communication that allow information to pass from observer to observer. And geometry, as measured in terms of area and volume, is nothing but a measure of the capacity of these screens to transmit information.

This more radical version of the holographic principle is based on the ideas introduced in Chapters 2 and 3. It relies strongly on the idea that the universe cannot be described from the point of view of an observer who exists somehow outside of it. Instead there are many partial viewpoints, where observers may receive information from their pasts. According to the holographic principle, geometrical quantities such as the areas of surfaces have their origins in measuring the flow of information to observers inside the universe.

Thus, it is not enough to say that the world is a hologram. The world must be a network of holograms, each of which contains coded within it information about the relationships between the others. In short, the holographic principle is the ultimate realization of the notion that the world is a network of relationships. Those relationships are revealed by this new principle to involve nothing but information. Any element in this network is nothing but a partial realization of the relationships between the other elements. In the end, perhaps, the history of a universe is nothing but the flow of information.

The holographic principle is still a new and very controversial idea. But for the first time in the history of quantum gravity we have in our hands an idea which at first seems too crazy to be true, but which survives all our attempts to disprove it. Whatever version of it finally turns out to be the true one, it is an idea which seems to be required by what we understand so far about quantum gravity. But it is also the kind of idea which will make it quite impossible, if it is ever accepted, to go back to any previous theory that did without it. The uncertainty principle of quantum theory and Einstein's equivalence principle were also ideas of this type. They contradicted the principles of older theories and, at first, just barely seemed to make sense. Just like them, the holographic principle is the kind of idea one hopes to run into just as one is turning the corner to a new universe.

....

CHAPTER 13

HOW TO WEAVE A STRING

Perhaps the main reason why some physicists do not get very excited about loop quantum gravity is that, although it succeeds very well in describing how the geometry of space must look on the Planck scale, it is basically pretty boring. There are no new principles involved. To set up the theory we just put in the basic principles of quantum theory and relativity. We get a lot out that is new and could even be tested experimentally. But it is perhaps not so surprising that when geometry is treated quantum theoretically it behaves like a quantum theoretic system. Things that used to be continuous, such as the range of possible volumes a space could have, now become discrete. The main lesson is that we really can treat space and time in a background independent way, and see them as a nothing but a network of relationships. This is good, but this is also what the principles we put in demanded. That it works is a good consistency check, but we should not consider it either surprising or revolutionary. The main strength of this approach, its simplicity and transparency, is perhaps also its main weakness.

String theory is just the opposite. We start not with basic principles, but by contradicting the thing we feel most certain about quantum gravity – that it must be a background independent theory. We ignore this, and search for a theory of gravitons and other particles moving against a background of empty space; and, by trial and error, we find it. Our guiding principle is to find something that works. To do this we have to change the rules, not once but over and over again. There are not particles, there are strings. There are not three dimensions of space but nine. There are extra symmetries. String theory is unique. Actually, it is not quite unique – it comes in an enormous number of versions. And in fact there are not just strings, but membranes of many different dimensions. And there are not nine dimensions, but ten. And so on. String theory has been nothing but a series of surprises, one after the other. We put in no principles – all we put in is the desire for a theory of gravitons that makes sense. And we get out a long list of unexpected facts, a whole new world to be explored.

For more than ten years, from about 1984 to 1996, these two theories of quantum gravity were developed by two different groups of people completely independently. Each group was successful in solving the problems it set for itself. Although we listened to each other's talks, and maintained friendships formed before the split, it must be said that almost everyone thought that their group was on the right path and the others were misguided. To each group it was obvious why the other could not succeed. The loop guys (and gals) said to the string guys, 'Your theory is not background independent, it cannot be a real quantum theory of space and time. Only we know how to make a successful background independent theory.' The string guys said to the loop guys, 'Your theory does not give a consistent description of the interactions between gravitons and other particles. Only our theory describes a consistent unification of gravity with the other interactions.' I am ashamed to admit that few in either community rose to the challenge. During this whole period, for example, there was not a single person who worked on both theories. Many seemed to make the understandable mistake of confusing the solution of part of the problem of quantum gravity with the solution of the whole problem.

Many misunderstandings have resulted. I have had the experience more than once of sitting next to someone from one camp listening to a talk by someone from the other. The person next to me would get very agitated: "That young person is so arrogant, they claim they have solved everything!" In fact the speaker had given a very measured presentation full of careful qualifications and caveats and had not made a single claim that went beyond what they had done. The problem is that such qualifications have to be presented in the terminology specific to the theory, and the person next to me, from the opposing theory's camp, was unable to follow it. This has happened to me in both directions. Even now, one can go to a conference and find that string theory and loop quantum gravity are the subjects of separate parallel sessions. The fact that the same problems are being addressed in the two sessions is noticed only by the small handful of us who do our best to be in both rooms.

There are many remarkable aspects of this situation, including the fact that almost every one of these people is quite sincere. Just as the existence of Moslems does not deter some Christians from the sincere conviction that theirs is the one true religion, and vice versa, there are many string theorists and many loop quantum gravity people who do not seem to be troubled by the existence of a whole community of equally sincere and smart people who pursue a different approach to the problem they are spending their lives attacking.

But this is a problem not of science but of the sociology of the academy. Sometimes, rushing from the loop room to the string room and back again, I have wondered what would have happened had physics in the seventeenth century been carried out in the same sociological context as present-day science. So let us wind back time and consider an alternative history of science. By 1630 there would have been two large groups of natural philosophers working on the successor to Aristotelian science. At conferences they would have divided into two parallel sessions with, as today, little overlap. In one room would be those who thought that falling bodies provided the key to the new physics. They would spend their time in profound reflections on the motion of bodies on the Earth. They would launch projectiles, experiment with pendulums and roll balls down inclined planes. Each of them would have their own personal version of the theory of falling bodies, but they would be united by the conviction that no theory could succeed that did not incorporate the deep principle discovered by Galileo that objects fall with a constant acceleration. They would be unconcerned with the motion of planets, because they would see nothing to disagree with the old and profoundly beautiful idea that planets move in circular orbits.

Two floors above them there would be a larger room where the ellipse theorists met. They would spend their time studying the orbits of planets, both in the real solar system and in imagined worlds of various dimensions. For them the key principle would be the great discovery by Kepler that planets move on elliptical orbits. They would be quite unconcerned with how bodies fall on Earth because they would share the view that only in the heavens could one see the true symmetries behind the world, uncontaminated by the complexities of the Earth, where so many bodies pushed on one another as they sought the centre. In any case they would be convinced that all motion, including that on Earth, must in the end reduce to complicated combinations of ellipses. They would assure sceptics that it was not yet time to study such problems, but when the time came they would have no problem explaining falling bodies in terms of the theory of ellipses.

Instead, they would focus their attention on the recent discovery of D-planets, which would have been found to follow parabolas rather than ellipses. So the definition of ellipse theory would be extended to include parabolas and other such curves such as hyperbolas. There would even be a conjecture that all the different orbits could be unified under one common theory, called C-theory. However, there was no agreed set of principles for C-theory, and most work on the subject required new mathematics that most physicists could not follow.

Meanwhile, another new form of mathematics was being invented by a brilliant mathematician and philosopher in Paris, René Descartes. He propounded a third theory, in which planetary orbits have to do with vortices.

It is true that while Galileo and Kepler did correspond, each seemed to show little interest in the key discoveries of the other. They wrote to each other about the telescope and what it revealed, but Galileo seems never to have mentioned ellipses, and to have gone to his grave believing the planetary orbits were circles. Nor is there any evidence that Kepler ever thought about falling bodies or believed them to be relevant to explaining the motions of the planets. It took a young scientist of a later generation, Isaac Newton, born the year of Galileo's death, to wonder whether the same force that made apples fall drew the Moon to the Earth and the planets to the Sun. So, while my story is fanciful, it really did happen that scientists with the stature of Galileo and Kepler each contributed an essential ingredient to a scientific revolution while remaining almost ignorant of and apparently uninterested in each other's discoveries.

We can hope that it will take less time to bring the different pieces of the quantum theory of gravity together than it did for someone to see the relationship between the work of Kepler and Galileo. The simple reason is that there are many more scientists working now than there were then. Whereas Kepler and Galileo might each have complained, if asked, that they were too busy to look at what the other was doing, there are now plenty of people to share the work. However, there is now the problem of making sure that young people have the freedom to wander across boundaries established by their elders without fear of jeopardizing their careers. It would be naive to say this is not a significant issue. In many areas of science we are paying for the consequences of an academic system that rewards narrowness of focus over exploration of new areas. This underlines the fact that good science is, and will always be, as much a question of judgement and character as it is a question of cleverness.

Indeed, over the last five years the climate of mutual ignorance and complacency that separated the string theorists from the loop quantum gravity people has begun to dissipate. The reason is that it has been becoming increasingly clear that each group has a problem it cannot solve. For string theory it is the problem of making the theory background independent and finding out what M theory really is. This is necessary both to unify the different string theories into a single theory and to make string theory truly a quantum theory of gravity. Loop quantum gravity is faced with the problem of how to show that a quantum spacetime described by an evolving spin network will grow into a large classical universe, which to a good approximation can be described in terms of ordinary geometry and Einstein's theory of general relativity. This problem arose in 1995 when Thomas Thiemann, a young German physicist then working at Harvard, presented for the first time a complete formulation of loop quantum gravity which resolved all the problems then known to exist. Thiemann's formulation built on all the previous work, to which he added some brilliant innovations of his own. The result was a complete theory which in principle should be able to answer any question. Furthermore, the theory could be derived directly from Einstein's general theory of relativity by following a well defined and mathematically rigorous procedure.

As soon as we had the theory, we began calculating with it. The first thing to calculate was how a graviton might appear as a description of a small wave or disturbance passing through a spin network. Before this could be done, however, we had to solve a more basic problem, which was to understand how the geometry of space and time, which seems so smooth and regular on the scales we can see, emerges from the atomic description in terms of spin networks. Until this was done we would not be able to make sense of what a graviton is, as gravitons should be related to waves in classical spacetime.

This kind of problem, new to us, is very familiar to physicists who study materials. If I cup my hands together and dip them into a stream I can carry away only as much water as will fill the 'cup'. But I can lift a block of ice just by holding it at its two sides. What is it about the different arrangements of the atoms in water and ice that accounts for the difference? Similarly, the spin networks that form the atomic structure of space can organize themselves in many different ways. Only a few of these ways will have a regular enough structure to reproduce the properties of space and time in our world.

What is remarkable – indeed, what is almost a miracle – is that the hardest problem faced by each group was precisely the key problem that the other had solved. Loop quantum gravity tells us how to make a background independent quantum theory of space and time. It offers a lot of scope to the *M* theorist looking for a way to make string theory background independent. On the other hand, if we believe that strings must emerge from the description of space and time provided by loop quantum gravity, we then have a lot of information about how to formulate the theory so that it does describe classical spacetime. The theory must be formulated in such a way that the gravitons appear not on their own, but as modes of excitations of extended objects that behave as strings.

It is then possible to entertain the following hypothesis: string theory and loop quantum gravity are each part of a single theory. This new theory will have the same relationship to the existing ones as Newtonian mechanics has to Galileo's theory of falling bodies and Kepler's theory of planetary orbits. Each is correct, in the sense that it describes to a good approximation what is happening in a certain limited domain. Each solves part of the problem. But each also has limits which prevent it from forming the basis for a complete theory of nature. I believe that this the most likely way in which the theory of quantum gravity will be completed, given the present evidence. In this penultimate chapter I shall describe some of this evidence, and the progress that has recently been made towards inventing a theory that unifies string theory and loop quantum gravity.

As a first step we can ask for a rough picture of how the two theories might fit together. As it happens, there is a very natural way in which strings and loops can emerge from the same theory. The key to this is a subtlety that I have so far only hinted at. Both loop quantum gravity and string theory describe physics on very small scales, roughly the Planck length. But the scale that sets the size of strings is not exactly equal to the Planck length. That scale is called the string length. The ratio of the Planck length to the string length is a number of great significance in string theory. It is a kind of charge, which tells us how strongly strings will interact with one another. When the string scale is much larger than the Planck length this charge is small and strings do not interact very much with one another.

We then can ask which scale is larger. There is evidence that, at least in our universe, the string scale is larger than the Planck scale. This is because their ratio determines the fundamental unit of electric charge, and that is itself a small number. We can then envisage scenarios in which loops are more fundamental. The strings will be descriptions of small waves or disturbances travelling through spin networks. Since the string scale is larger, we can explain the fact that string theory relies on a fixed background, as the needed background can be supplied by a network of loops. The fact that strings seem to experience the background as a continuous space is explained by them being unable to probe down to a distance where they can distinguish a smooth background from a network of loops (see Figure 38 on page 165).

One way to talk about this is that space may be 'woven' from a network of loops, as shown in Figure 38, just as a piece of cloth is woven from a network of threads. The analogy is fairly precise. The properties of the cloth are explicable in terms of the kind of weave, which is to say in terms of how the threads are knotted and linked with one another. Similarly, the geometry of the space we may weave from a large spin network is determined only by how the loops link and intersect one another.

We may then imagine a string as a large loop which makes a kind of embroidery of the weave. From a microscopic point of view, the string can be described by how it knots the loops in the weave. But on a larger scale we would see only the loop making up the string. If we cannot see the fine weave that makes up space, the string will appear against a background of some apparently smooth space. This is how the picture of strings against a background space emerges from loop quantum gravity.

If this is right, then string theory will turn out to be an approximation to a more fundamental theory described in terms of spin networks. Of course, just because we can argue for a picture like this does not mean that it can be made to work in detail. In particular, it may not work for any version of loop quantum gravity. To make the large loops behave as strings we may have to choose the details of the loop theory carefully. This is good, not bad, for it tells us how information about the world already revealed by string theory may be coded in such a way that it becomes part of the fundamental theory that describes the atomic structure of space and time. At present, a programme of research is under way to unify string theory and loop quantum gravity using essentially this idea. Very recently this has led to the discovery of a new theory that appears to contain within it both string theory and a form of loop quantum gravity. It looks promising to some of us but, as it is work in progress, I can say no more about it here.

However, if this programme does work it will exactly realize the idea of duality I discussed in Chapter 9. It will also realize the aims of Amitaba Sen, for the whole loop approach arose out of his efforts to understand how to quantize supergravity, which is now understood to be closely related to string theory.

While my hypothesis is certainly not proven, evidence has been accumulating that string theory and loop quantum gravity may describe the same world. One piece of evidence, discussed in the last chapter, is that both theories point to some version of the holographic principle. Another is that the same mathematical ideas structures keep appearing on both sides. One example of this is a structure called *non-commutative geometry*. This is an idea about how to unify quantum theory with relativity that was invented by the French mathematician Alain Connes. The basic idea is very simple: in quantum physics we cannot measure the position and velocity of a particle at the same time. But if we want to we can at least determine the position precisely. However, notice that a determination of the position of a particle actually involves three different measurements, for we must measure where the particle is relative to a set of three axes (these measurements yield the three components of the *position vector*). So we may consider an extension of the uncertainty principle in which one can measure only one of these components precisely at any one time. When it is impossible to measure two quantities simultaneously, they are said not to commute, and this idea leads to a new kind of geometry which is labelled non-commutative. In such a world one cannot even define the idea of a point where something may be exactly located.

Alain Connes's non-commutative geometry thus gives us another way to describe a world in which the usual notion of space has broken down. There are no points, so it does not even make sense to ask if there are an infinite number of points in a given region. What is really wonderful, though, is that Connes has found that large pieces of relativity theory, quantum theory and particle physics can be carried over into such a world. The result is a very elegant structure that seems also to penetrate to several of the deepest problems in mathematics.

At first, Connes's ideas were developed independently of the other approaches. But in the last few years people have been surprised to discover that both loop quantum gravity and string theory describe worlds in which the geometry is non-commutative. This gives us a new language in which to compare the two theories.

One way to test the hypothesis that strings and loops are different ways of describing the same physics is to attack a single problem with both methods. There is an obvious target: the problem of giving a description of a quantum black hole. From the discussion in Chapters 5 to 8, we know that the main objective is to explain in terms of some fundamental theory where the entropy and temperature of a black hole come from, and why the entropy is proportional to the area of the black hole's horizon. Both string theory and loop quantum gravity have been used to study quantum black holes, with spectacular results coming on each side in the last few years.

The main idea on each side is the same. Einstein's theory of general relativity is to be thought of as a macroscopic description, obtained by averaging over the atomic structure of spacetime, in exactly the same way that thermodynamics is obtained by applying statistics to the motion of atoms. Just as a gas is described roughly in terms of continuous quantities such as density and temperature, with no mention of atoms, in Einstein's theory space and time are described as continuous, and no mention is made of the discrete, atomic structure that may exist on the Planck scale.

Given this general picture, it is natural to ask whether the black hole's entropy is a measure of the missing information that could be obtained from an exact quantum description of the geometry of space and time around a black hole. The fact that the entropy of a black hole is proportional to the area of its horizon should be a huge clue to its meaning. String theory and loop quantum gravity have each found a way to use this clue to construct a description of a quantum black hole.

In string theory, good progress has been made by conjecturing that the missing information measured by the black hole's entropy is a description of how the black hole was formed. A black hole is a very simple object. Once formed, it is featureless. From the outside one can measure only a few of its properties: its mass, electric charge and angular momentum. This means that a particular black hole might have been formed in many different ways: for example, from a collapsing star, or – in theory at least – by compressing, say, a pile of science-fiction magazines to an enormous density. Once the black hole has formed there is no way to look inside and see how it was formed. It emits radiation, but that radiation is completely random, and offers no clue to the black hole's origin. The information about how the black hole formed is trapped inside it. So one may hypothesize that it is exactly this missing information that is measured by the black hole's entropy.

Over the last few years string theorists have discovered that string theory is not just a theory of strings. They have found that the quantum gravity world must be full of new kinds of object that are like higher-dimensional versions of strings in that they extend in several dimensions. Whatever their dimension, these objects are called *branes*. This is shortened from 'membranes', the term used for objects with two spatial dimensions. The branes emerged when new ways to test the consistency of string theory were discovered, and it was found that the theory can be made mathematically consistent only by including a whole set of new objects of different dimensions.

String theorists have found that in certain very special cases black holes could be made by bringing together a collection of these branes. To do this they make use of a feature of string theory, which is that the gravitational force is adjustable. It is given by the value of a certain physical field. When this field is increased or decreased, the gravitational force becomes stronger or weaker. By adjusting the value of the field it is possible to turn the gravitational force on and off. To make a black hole they begin with the gravitational field turned off. Then they imagine assembling a set of branes which have the mass and charge of the black hole they want to make. The object is not yet a black hole, but they can turn it into one by turning up the strength of the gravitational force. When they do so a black hole must form.

String theorists have not yet been able to model in detail the process of the formation of the black hole. Nor can they study the quantum geometry of the resulting black hole. But they can do something very cute, which is to count the number of different ways that a black hole could be formed in this way. They then assume that the entropy of the resulting black hole is a measure of this number. When they do the counting, they get, right on the nose, the right answer for the entropy of the black hole.

So far only very special black holes can be studied by this method. These are black holes whose electric charges are equal to their mass. This is to say that the electrical repulsions of two of these black holes are exactly balanced by their gravitational attractions. As a result, one can put two of them next to each other and they will not move, for there is no net force between them. These black holes are very special because their properties are strongly constrained by the condition that their charge balances their mass. This makes it possible to get precise results, and, when this is possible the results are very impressive. On the other hand, it is not known how to extend the method to all black holes. Actually, string theorists can do a bit better than this, for the methods can be used to study black holes whose charges are close to their masses. These calculations also give very impressive results: in particular, they reproduce every last factor of 2 and

π in the formula for the radiation emitted by these black holes.

A second idea about a black hole's entropy is that it is a count, not of the ways to make a black hole, but of the information present in an exact description of the horizon itself. This is suggested by the fact that the entropy is proportional to the area of the horizon. So the horizon is something like a memory chip, with one bit of information coded in every little pixel, each pixel taking up a region 2 Planck lengths on a side. This picture turns out to be confirmed by calculations in loop quantum gravity.

A detailed picture of the horizon of a black hole has been developed using the methods of loop quantum gravity. This work started in 1995 when, inspired by the ideas of Crane, 't Hooft, and Susskind, I decided to try to test the holographic principle in loop quantum gravity. I developed a method for studying the quantum geometry of a boundary or a screen. As I mentioned earlier, the result was that the Bekenstein bound was always satisfied, so that the information coded into the geometry on the boundary was always less than a certain number times its total area.

Meanwhile, Carlo Rovelli was developing a rough picture of the geometry of a black hole horizon. A graduate student of ours, Kirill Krasnov, showed me how the method I had discovered could be used to make Carlo's ideas more precise. I was quite surprised because I had thought that this would be impossible. I worried that the uncertainty principle would make it impossible to locate the horizon exactly in a quantum theory. Kirill ignored my worries and developed a beautiful description of the horizon of a black hole which explained both its entropy and its temperature. (Only much later did Jerzy Lewandowksi, a Polish physicist who has added much to our understanding of loop quantum gravity, work out how the uncertainty principle is circumvented in this case.)

Kirill's work was brilliant, but a bit rough. He was subsequently joined by Abhay Ashtekar, John Baez, Alejandro Corichi and other more mathematically minded people who developed his insights into a very beautiful and powerful description of the quantum geometry of horizons. The results can be applied very widely, and give a general and completely detailed description of what a horizon would look like were it to be probed on the Planck scale.

While this work applies to a much larger class of black holes than can be addressed by string theory, it does have one shortcoming compared with string theory: there is one constant that has to be adjusted to make the entropy and temperature come out right. This constant determines the value of Newton's gravitational constant, as measured on large scales. It turns out that there is a small change in the value of the constant when one compares its value measured on the Planck scale with the value measured at large distances. This is not surprising. Shifts like this occur commonly in solid state physics, when one takes into account the effect of the atomic structure of matter. This shift is finite, and has to be made just once, for the whole theory. (It is actually equal to the $\sqrt{3}/\log 2$.) Once done it brings the results for all different kinds of black holes in exact agreement with the predictions by Bekenstein and Hawking that we discussed in Chapters 6 to 8.

Thus, string theory and loop quantum gravity have each added something essential to our understanding of black holes. One may ask whether there is a conflict between the two results. So far none is known, but this is largely because, at the moment, the two methods apply to different kinds of black hole. To be sure, we need to find a way of extending one of the methods so that it covers the cases covered by the other method. When we can do this we will be able to make a clean test of whether the pictures of black holes given by loop quantum gravity and string theory are consistent with each other.

This is more or less what we have been able to understand so far about black holes from the microscopic point of view. A great deal has been understood, although it must also be said that some very important questions remain unanswered. The most important of these have to do with the interiors of black holes. Quantum gravity should have something to say about the singular region in the interior of a black hole, in which the density of matter and the strength of the gravitational field become infinite. There are speculations that quantum effects will remove the singularity, and that one consequence of this may be the birth of a new universe inside the horizon. This idea has been studied using approximation techniques in which the matter forming the black hole is treated quantum theoretically, but the geometry of spacetime is treated as in the classical theory. The results do suggest that the singularities are eliminated, and one may hope that this will be confirmed by an exact treatment. But, at least so far, neither string theory nor loop quantum gravity, nor any other approach, has been strong enough to study this problem.

Until 1995 no approach to quantum gravity could describe black holes in any detail. None could explain the meaning of the entropy of a black hole or tell us anything about what black holes look like when probed on the Planck scale. Now we have two approaches that are able to do all these things, at least in some cases. Every time we are able to calculate something about a black hole, in either theory, it comes out right. There are many questions we still cannot answer, but it is difficult to avoid the impression that we are finally understanding something real about the nature of space and time.

Furthermore, the fact that both string theory and loop quantum gravity both succeed in giving the right answers about quantum black holes is strong evidence that the two approaches may be revealing different sides of a single theory. Like Galileo's projectiles and Kepler's planets, there is more and more evidence that we are glimpsing the same world through different windows. To find the relation of his work to Kepler's, Galileo would only have had to imagine throwing a ball far enough and fast enough that it became a moon. Kepler, from his point of view, could have imagined what a planet orbiting very close to the Sun might have looked like to people living on the Sun. In the present case, we only have to ask whether a string can be woven from a network of loops, or whether, if we look closely enough at a string, we can see the discrete structures of the loops. I personally have little doubt that in the end loop quantum gravity and string theory will be seen as two parts of a single theory. Whether it will take a Newton to find that theory, or whether it is something we mortals can do, is something that only time will tell.

....

CHAPTER 14

WHAT CHOOSES THE LAWS OF NATURE?

Back in the 1970s there was a simple dream about how physics would end. A unified theory would be found that incorporated quantum theory, general relativity, and the various particles and forces known to us. This would not only be a theory of everything, it would be unique. We would discover that there was only one mathematically consistent quantum theory that unified elementary particle physics with gravity. There could be only one right theory and we would have found it. Because it was unique, this theory would have no free parameters – there would be no adjustable masses or charges. If there was anything to adjust, the theory would then come in different versions, and it would not be unique. There would be only one scale, against which everything was measured, which was the Planck scale. The theory would allow us to calculate the results of any experiment to whatever accuracy we desired. We would calculate the masses of the electron, proton, neutron, neutrinos and all the other particles, and our results would all be in exact agreement with experiments.

These calculations would have to explain certain very strange features of the observed masses of the particles. For example, why are the masses of the proton and neutron so very small in Planck units? Their masses are of the order of 10^{-19} Planck masses. Where do such terribly small numbers come from? How could they come out of a theory with no free parameters? If the fundamental atoms of space have the

Planck dimensions, then we would expect the elementary particles to have similar dimensions. The fact that protons and neutrons are nearly 20 orders of magnitude lighter than the Planck mass seems very hard to understand. But since the theory would be unique it would have to get this right.

String theory was invented with the hope that it would be this final theory. It was its potential uniqueness that made it worth studying, even as it became clear that it was not soon going to lead to predictions about the masses of particles or anything else that could be tested experimentally. After all, if there is one unique theory it does not need experiments to verify it – all that is needed is to show that it is mathematically consistent. A unique theory must automatically be proved right by experiments, so it does not matter if a test of the theory is several centuries away. If we accept the assumption that there is one unique theory, then it will pay to concentrate on the problem of testing that theory for mathematical consistency rather than on developing experimental tests for it.

The problem is that string theory did not turn out to be unique. It was instead found to come in a very large number of versions, each equally consistent. From our present-day perspective, taking into account only the results on the table, it seems that the hope for a unique theory is a false hope. In the current language of string theory, there is no way to distinguish between any of a very large number of theories: they are all equally consistent. Moreover, many of them have adjustable parameters, which could be changed to obtain agreement with experiment.

Looking back, it is clear that the assumption that a unified theory would be unique was no more than that – an assumption. There is no mathematical or philosophical principle which guarantees there to be only one mathematically consistent theory of nature. In fact, we now know that there can be no such theory. For example, suppose that the world had one or two spatial dimensions, rather than three. For these cases we have constructed lots of consistent quantum theories, including some which have gravity. These were done as warm-up exercises for various research programmes. We keep them around as experimental laboratories in which we can test new ideas in a context where we know we can calculate anything we like. It is always possible that there is only one possible consistent theory to describe worlds that have more than two spatial dimensions. But there is no known reason why this should be true. In the absence of any evidence to the contrary, the fact that there are many consistent theories that describe one- and two-dimensional universes should lead us to doubt the assumption that mathematical consistency in itself allows only one theory of nature.

Of course, there is a way out, which is the possibility that string theory is not the final theory. Besides the fact that it comes in many versions, there are good reasons to believe this: string theory is background dependent and it is understood only in terms of a certain approximation scheme. A fundamental theory needs to be background independent and capable of being formulated exactly. So most people who work with string theory now believe the M theory conjecture I described in Chapter 11: that there is a single theory, which can be written down exactly and in a way that is independent of any given spacetime, that unifies all the different string theories.

There is some evidence to support this M theory conjecture. Many physicists, myself included, are now trying to invent the theory. There seem to be three possibilities:

1 The correct theory of nature is not a string theory.

2 The M theory conjecture is false: there is no unified string theory, but one of the string theories will make predictions that agree with experiment.

3 The M theory conjecture is true: there is a single unified theory, which, however, predicts that the world could come in a great many different physical phases. In these phases the laws of nature appear to be different. Our universe is in one of them.

If possibility 1 is true, then all we can do is take the story of string theory as a cautionary tale. So let us put this one aside and look at the others. If possibility 2 is true, then we are left with a puzzle: what or who chose which consistent theory applies to our world? Among the list of different possible consistent theories, how was one chosen to apply to our universe?

There seems to be only one possible answer to this question. Something external to the universe made the choice. If that's the way things turn out, then this is the exact point at which science will become religion. Or, to put it better, it will then be rational to use science as an argument for religion. One already hears a lot about this in theological circles, as well as from certain scientists, in the form of arguments based on what we might call the *anthropic observation*. It seems that the universe we live in is very special. For a universe to exist for billions of years and contain the ingredients for life, certain special conditions must be satisfied: the masses of the elementary particles and the strengths of the fundamental forces must be tuned to values very close to the ones actually we observe. If these parameters are outside certain narrow limits, the universe will be inhospitable to life. This raises a legitimate scientific question: given that there seem to be more than one possible consistent set of laws, why is it that the laws of nature are such that the parameters fall within the narrow ranges needed for life? We may call this the *anthropic question*.

If there are different possible consistent laws of nature, but no framework which unifies them, then there are only two possible answers to the anthropic question. The first is that we are very lucky indeed. The second is that whatever entity specified the laws did so in order that there would be life. In this case we have an argument for religion. This is of course a version of an argument which is well known to theologians – the God of the Gaps argument. If science raises a question like the anthropic question that cannot be answered in terms of processes that obey the laws of nature, it becomes rational to invoke an outside agency such as God. The scientific version of this argument is called the *strong anthropic principle*.

Notice that this argument is valid only if there is no way to explain how the laws of nature might have been chosen except by invoking the action of some entity outside our universe. You may recall the principle with which I started this book: that there is nothing outside the universe. As long as there is a way of answering all our questions without violating this principle, we are doing science and we have no need of any other mode of explanation. So the argument for the strong anthropic principle has logical force only if there is no other possibility.

But there is another possibility, possibility 3. This is like possibility 2, but with an important difference. If the different string theories describe different phases of a single theory, then it is possible that under the right circumstances there could be a transition from one phase to another. Just as ice melts to water, the universe could 'melt' from one phase, in which it is described by one string theory, to another phase, in which it is described by another. We are then still left with the question of why one phase rather than another describes our universe, but this is not so hard to resolve because in this picture the universe is allowed to have changed phase as it evolved in time. There is also the possibility that different regions of the universe exist in different phases.

Given these possibilities, there are at least two alternatives to the God of the Gaps argument. The first is that there is some process that creates many universes. (Do not worry for the moment about what that process is, for cosmologists have found several attractive ways to make a universe which continually spawns new universes.) The big bang is then not the origin of all that exists, but only a kind of phase transition by which a new region of space and time was created, in a phase different than the one from which it came, and then cooled and expanded. In such a scenario there could be many big bangs, leading to many universes. The astrophysicist Martin Rees has a nice name for this – he calls the whole collection the 'multiverse'. It is possible that the process creates universes in random phases. Each would then be governed by a different string theory. These universes will have different dimensions and geometries, and they will also have different sets of elementary particles which interact according to different sets of laws. If there are adjustable parameters, it is possible that they are set at random each time a new universe is created.

So there is a simple answer to the anthropic question. Among all the possible universes, a minority will have the property that their laws are hospitable to life. Since we are alive, we naturally find ourselves in one of them. And since there are a great many universes, we need not worry that the chance of any one of them being hospitable to life is small, because the chance of at least one of them being hospitable to life may not be small. There will then be nothing to explain. Martin Rees likes to put this in the following way: if one finds a bag by the side of the road containing a suit that fits one perfectly, that is something to wonder about. But if one goes into a clothing store and is able to find a suit that fits, there is no mystery because the store carries lots of suits in many different sizes. We may call this the God of The Gap. It is also sometimes called the *weak anthropic principle*.

The only problem with this kind of explanation is that it is difficult to see how it could be refuted. As long as your theory yields a very large number of universes, you only need there to be at least one like ours. The theory makes no other predictions apart from the existence of at least one universe like ours. But we already know that, so there is no way to refute this theory. This might seem good, but actually it is not because a theory that cannot be refuted cannot really be part of science. It can't carry very much explanatory weight, because whatever features our universe has, as long as it can be described by one of the large number of string theories, our theory will not be refuted. Therefore it can make no new predictions about our universe.

Is it possible to have a theory which gives a scientific answer to the anthropic question? Such a theory may be framed around the possibility that the universe can make a physical transition from one phase to another. If we could look back into the history of the universe to before the big bang, it may be that we would see one or a whole succession of different phases in which the universe had different dimensions and appeared to satisfy different laws. The big bang would then be just be the most recent of a series of transitions the universe has passed through. And even though each phase may be governed by a different string theory, the whole history of the universe would be governed by a single law – M theory. We then need an explanation in physical terms for why the universe 'chose' to exist in a phase such as the one in which we find ourselves, which exists for billions of years and is hospitable for life. There are several different possible explanations of this kind, which are described in detail in another book of mine, *The Life of the Cosmos*, so I shall be brief here.

One idea is that new universes could form inside black holes. In this case our universe would have a large number of progeny, as it contains at least 10^{18} black holes. One may also conjecture that the changes in the laws from old universes to new are small, so that the laws in each new universe formed from our own are close to those that hold in our universe. This also means that the laws in the universe from which ours was formed were not very different from those of our own. Given these two assumptions, a mechanism which is formally analogous to natural selection operates, because after many generations those universes that give rise to many black holes will dominate the population of universes. The theory then predicts that a randomly chosen universe will have the property that it will make more black holes than will universes with slightly different values of the parameters. We can then ask whether this prediction is satisfied by our universe. To cut a long story short, up to the present time it seems that it is. The reason is that carbon chemistry is not only good for life, it plays an important role in the processes that make the massive stars that end up as black holes. However, there are several possible observations which could disprove the theory. Thus, unlike the God of the Gaps and the God of The Gap theories, this theory is very vulnerable to being disproved. Of course, this means that it is likely to be disproved.

The important thing about this theory is that it shows that there are alternatives to both the strong and the weak anthropic principle. And if that is so, then those principles have no logical force. Beyond this, the theory of *cosmological natural selection* (as it is sometimes called) shows us that physics can learn an important lesson from biology about possible modes of scientific explanation. If we want to stick to our principle that there is nothing outside the universe, then we must reject any mode of explanation in which order is imposed on the universe by an outside agency. Everything about the universe must be explicable only in terms of how the laws of physics have acted in it over the whole span of its history.

Biologists have been facing up to this problem for more than a century and a half, and they have understood the power of different kinds of mechanism by which a system may organize itself. These include natural selection, but that is not the only possibility: other mechanisms of self-organization have been discovered more recently. These include self-organized critical phenomena, invented by Per Bak and collaborators and studied by many people since. Other mechanisms of self-organization have been studied by theoretical biologists such as Stuart Kauffman and Harold Morowitz. So there is no shortage of mechanisms for self-organization that we could consider in this context. The lesson is that if cosmology is to emerge as a true science, it must suppress its instinct to explain things in terms of outside agencies. It must seek to understand the universe on its own terms, as a system that has formed itself over time, just as the Earth's biosphere has formed itself over billions of years, starting from a soup of chemical reactions.

It may seem fantastic to think of the universe as analogous to a biological or ecological system, but these are the best examples we have of the power of the processes of self-organization to form a world of tremendous beauty and complexity. If this view is to be taken seriously, we should ask whether there is any evidence for it. Are there any aspects of the universe and the laws that govern it that require explanation in terms of mechanisms of self-organization? We have already discussed one piece of evidence for this, which is the anthropic observation: the apparently improbable values of the masses of the elementary particles and the strengths of the fundamental forces. One can estimate the probability that the constants in our standard theories of the elementary particles and cosmology would, were they chosen randomly, lead to a world with carbon chemistry. That probability is less than one part in 10^{220}. But without carbon chemistry the universe would be much less likely to form large numbers of stars massive enough to become black holes, and life would be very unlikely to exist. This is evidence for some mechanism of self-organization, because what we mean by self-organization is a system that evolves from a more probable to a less probable configuration. So the best argument we can give that such a mechanism has operated in the past must have two parts: first, that the system be structured in some way that is enormously improbable; and second, that nothing acting from the outside could have imposed that organization on the system. In the case of our universe we are taking this second part as a principle. We then satisfy both parts of the argument, and are justified in seeking mechanisms of self-organization to explain why the constants in the laws of nature have been chosen so improbably.

But there is an even better piece of evidence for the same conclusion. It is right in front of us, and so familiar that it is difficult at first to understand that it also is a structure of enormous improbability. This is space itself. The simple fact that the world consists of a three-dimensional space, which is almost Euclidean in its geometry, and which extends for huge distances on all sides, is itself an extraordinarily improbable circumstance. This may seem absurd, but this is only because we have become so mentally dependent on the Newtonian view of the world. For how probable the arrangement of the universe is cannot be answered a priori. Rather, it depends on the theory we have about what space is. In Newton's theory we posit that the world lives in an infinite three-dimensional space. On this assumption, the probability of us perceiving a three-dimensional space around us, stretching infinitely in all directions, is 1. But of course we know that space is not exactly Euclidean, only approximately so. On large scales space is curved because gravity bends light rays. Since this directly contradicts a prediction of Newton's theory, we can deduce that, with probability 1, Newton's theory is false.

It is a little harder to pose the question in Einstein's theory of spacetime, as that theory has an infinite number of solutions. In many of them space is approximately flat, but in many of them it is not. Given that there are an infinite number of examples of each, it is not straightforward to ask how probable it would be, were the solution chosen at random, that the resulting universe would look almost like three-dimensional Euclidean space.

It is easier to ask the question in a quantum theory of gravity. To ask it we need a form of the theory that does not assume the existence of any classical background geometry for space. Loop quantum gravity is an example of such a theory. As I explained in Chapters 9 and 10, it tells us that there is an atomic structure to space, described in terms of the spin networks invented by Roger Penrose. As we saw there, each possible quantum state for the geometry of space can be described as a graph such as that shown in Figures 24 to 27. We can then pose the question this way: how probable is it that such a graph represents a geometry for space that would be perceived by observers like us, living on a scale hugely bigger than the Planck scale, to be an almost Euclidean three-dimensional space? Well, each node of a spin network graph corresponds to a volume of roughly the Planck length on each side. There are then 10^{99} nodes inside every cubic centimetre. The universe is at least 10^{27} centimetres in size, so it contains at least 10^{180} nodes. The question of how probable it is that space looks like an almost flat Euclidean three-dimensional space all the way up to cosmological scales can then be posed as follows: how probable is it that a spin network with 10^{180} nodes would represent such a flat Euclidean geometry?

The answer is, exceedingly improbable! To see why, an analogy will help. To represent an apparently smooth, feature-less three-dimensional space, the spin network has to have some kind of regular arrangement, something like a crystal. There is nothing special about any position in Euclidean space that distinguishes it from any other position. The same must be true, at least to a good approximation, of the quantum description of such a space. Such a spin network must then be something like a metal. A metal looks smooth because the atoms in it have a regular arrangement, consisting of almost perfect crystals that contain huge numbers of atoms. So the question we are asking is analogous to asking how probable it is that all the atoms in the universe would arrange themselves in a crystalline structure like the atoms in a metal, stretching from one end of the universe to another. This is, of course, exceedingly improbable. But there are about 10^{75} spin network nodes inside every atom, so the probability that all of them are arranged regularly is less than 1 part in 10^{75} – smaller still.

It may be that this is an underestimate and the probability is not quite so small. There is one way of ensuring that all the atoms in the universe arrange themselves in a perfect crystal, which is to freeze the universe down to a temperature of absolute zero, and compacted so as to give it a density high enough for hydrogen gas to form a solid. So perhaps the spin network representing the geometry of the world is arranged regularly because it is frozen.

We can ask how probable this is. We can reason that if the universe were formed completely by chance it would have a temperature which is some reasonable fraction of the maximum possible temperature. The maximum possible temperature is the temperature that a gas would have if each atom was as massive as the Planck mass and moved at a fair fraction of the speed of light. The reason is that if the temperature were raised beyond point, the Planck temperature, the molecules would all collapse into black holes. Now, for the atoms of space to have a regular arrangement the temperature must be much, much less than this maximum temperature. In fact, the temperature of the universe is less than 10^{-32} times the Planck temperature. So the probability that a universe, chosen randomly, would have this temperature is less than 1 part in 10^{32}. So we conclude that it is at least this improbable that the universe is as cold as it is.

Whichever way we make the estimate, we conclude that if space really has a discrete atomic structure, then it is extraordinarily improbable that it would have the completely smooth and regular arrangement we observe it to have. So this is indeed something that requires explanation. If the explanation is not to be that some outside agency chose the state of the universe, there must have been some mechanism of self-organization that, acting in our past, drove the world into this incredibly improbable state. Cosmologists have been worrying about this problem for some time. One solution which has been proposed is called *inflation*. This is a mechanism by which the universe can blow itself up exponentially fast until it becomes the flat, almost Euclidean universe we observe today. Inflation solves part of the problem, but it itself requires certain improbable conditions. When inflation begins to act, the universe must already be smooth on a scale of at least 10^5 times the Planck scale. And – at least as far as we know – inflation requires the fine tuning of two parameters. One of these is the cosmological constant, which must be smaller than its natural value in a quantum theory of gravity by a factor of at least 10^{60}. The other is the strength of a certain force, which in many versions of inflation must be no greater than 10^{-6}. The net result is that for inflation to act we require a situation with a probability of at most 10^{-81}. Even if we leave the cosmological constant out of it, we still require a situation whose probability is at most 10^{-21}. So inflation may be part of the answer, but it cannot be the complete answer.

Is it possible that some method of self-organization accounts for the fact that space looks perfectly smooth and regular, on scales hugely bigger than the Planck scale? This question has prompted some recent research, but as yet no clear answer has emerged. But if we are to avoid an appeal to religion, then this is a question that must have an answer.

So, in the end, the most improbable and hence the most puzzling aspect of space is its very existence. The simple fact that we live in an apparently smooth and regular three-dimensional world represents one of the greatest challenges to the developing quantum theory of gravity. If you look around at the world seeking mystery, you may reflect that one of the biggest mysteries is that we live in a world in which it is possible to look around, and see as far as we like. The great triumph of the quantum theory of gravity may be that it will explain to us why this is so. If it does not, then the mystic who said that God is all around us will turn out to have been right. But if we find a scientific explanation of the existence of space, and so take the wind out of the sails of such a theistic mystic, there will still remain the mystic who preaches that God is nothing but the power of the universe as a whole to organize itself. In either case the greatest gift the quantum theory of gravity could give the world would be a renewed appreciation of the miracle that the world exists at all, together with a renewed faith that at least some small aspect of this mystery may be comprehended.

If I have done my job well I shall have left you with an understanding of the questions being asked by those of us who are aiming to complete the twentieth-century revolution in physics. One or another, or several, or none of the theories I have discussed may turn out to be right, but I hope that you will at least have gained an appreciation of what is at stake and what it will mean when we do finally find the quantum theory of gravity. My own view is that all the ideas I have discussed here will turn out to be part of the picture – that is why I have included them. I hope I have been sufficiently clear about my own views for you to have had no trouble distinguishing them from well established parts of science such as quantum theory and general relativity.

But above all, I hope you will have been persuaded that the search for fundamental laws and principles is one that is well worth supporting. For our community of researchers depends totally on the community at large for support in our endeavour. This reliance is twofold. First, it matters to us a great deal that we are not the only ones who care what space and time are, or where the universe came from. While I was writing my first book, I worried a lot over the time I was spending not doing science. But I found instead that I gained tremendous energy from all the interactions with ordinary people who take the time to follow what we do. Others I have spoken with have had the same experience. The most exciting thing about being in the position of conveying the cutting edge of science to the public is discovering how many people out there care whether we succeed or fail in our work. Without this feedback there is a danger of becoming stale and complacent, and seeing our contributions only in terms of the narrow criteria of academic success. To avoid this we have to keep alive the feeling that our work brings us into contact with something true about nature. Many young scientists have this feeling, but in today's competitive academic environment it is not easy to maintain it over a lifetime of research. There is perhaps no better way to rekindle this feeling than to communicate with people who bring to the conversation nothing more than a strong desire to learn.

The second reason why we depend on the public for support is that most of us produce nothing but this work. Since we have nothing to sell, we depend on the generosity of society to support our research. This kind of research is inexpensive, compared with medical research or experimental elementary particle physics, but this does not make it secure. The present-day political and bureaucratic environment in which science finds itself favours big, expensive science – projects that bring in the level of funding that boosts the careers of those who make the decisions about which kinds of science get supported. Nor is it easy for responsible people to commit funds to a high-risk field like quantum gravity, which has so far no experimental support to show for it. Finally, the politics of the academy acts to decrease rather than increase the variety of approaches to any problem. As more positions become earmarked for large projects and established research programs, there are correspondingly fewer positions available for young people investigating their own ideas. This has unfortunately been the trend in quantum gravity in recent years. This is not deliberate, but it is a definite effect of the procedures by which funding officers and deans measure success. Were it not for the principled commitment of a few funding officers and a few departmental heads and, not the least, a few private foundations, this kind of fundamental, high-risk/high-payoff research would be in danger of disappearing from the scene.

And quantum gravity is nothing if not high risk. The unfortunate lack of experimental tests means that relatively large groups of people may work for decades only to find that they have completely wasted their time, or at least done little but eliminate what at first seemed to be attractive possibilities for the theory. Measured sociologically, string theory seems very healthy at the moment, with perhaps a thousand practitioners; loop quantum gravity is robust but much less populous, with about a hundred investigators; other directions, such as Penrose's twistor theory, are still pursued by only a handful. But thirty years from now all that will matter is which parts of which theory were right. And a good idea from one person is still worth hundreds of people working incrementally to advance a theory without solving its fundamental problems. So we cannot allow the politics of the academy too much influence here, or we shall all end up doing one thing. If that happens, then a century from now people may still be writing books about how quantum gravity is almost solved. If this is to be avoided, all the good ideas must be kept alive. Even more important is to maintain a climate in which young people feel there is a place for their ideas, no matter how initially unlikely or how far from the mainstream they may seem. As long as there is still room for the young scientist with the uncomfortable question and the bright idea, I see nothing to prevent the present rapid rate of progress from continuing until we have a complete theory of quantum gravity.

I should like to close this book by sticking out any part of my neck which is not yet exposed, and making a few predictions about how the problem of quantum gravity will in the end be solved. I believe that the huge progress we have made in the last twenty years is best illustrated by the fact that it is now possible to make an educated guess about how the last stages of the search for quantum gravity will go. Until recently we could have done no more than point to a few good ideas that were not obviously wrong. Now we have several proposals on the table that seem right enough and robust enough, and it is hard to imagine that they are completely wrong. The picture I have presented in this book was assembled by taking all those ideas seriously. In the same spirit, I offer the following scenario of how the present revolution in physics will end.

- Some version of string theory will remain the right description at the level of approximation at which there are quantum objects moving against a classical spacetime background. But the fundamental theory will look nothing like any of the existing string theories.

- Some version of the holographic principle will turn out to be right, and it will be one of the foundational principles of the new theory. But it will not be the strong version of the principle I discussed in Chapter 12.

- The basic structure of loop quantum gravity will provide the template for the fundamental theory. Quantum states and processes will be expressed in diagrammatic form, like the spin networks. There will be no notion of a continuous geometry of space or spacetime, except as an approximation. Geometrical quantities, including areas and volumes, will turn out to be quantized, and to have minimum values.

- A few of the other approaches to quantum gravity will turn out to play significant roles in the final synthesis. Among them will be Roger Penrose's twistor theory and Alain Connes's non-commutative geometry. These will turn out to give essential insights into the nature of the quantum geometry of spacetime.

- The present formulation of quantum theory will turn out to be not fundamental. The present quantum theory will first give way to a relational quantum theory of the kind I discussed in Chapter 3, which will be formulated in the language of topos theory. But after a while this will be reformulated as a theory about the flow of information among events. The final theory will be non-local or, better, extra-local, as space itself will come to be seen only as an appropriate description for certain kinds of universe, in the same way that thermodynamic quantities such as heat and temperature are meaningful only as averaged descriptions of systems containing many atoms. The idea of 'states' will have no place in the final theory, which will be framed around the idea of processes and the information conveyed between them and modified within them.

- Causality will be a necessary component of the fundamental theory. That theory will describe the quantum universe in terms of discrete events and their causal relations. The notion of causality will survive at a level in which space will no longer be a meaningful concept.

- The final theory will not be able to predict unique values for the masses of the elementary particles. The theory will allow a set of possible values for these and other quantities in fundamental physics. But there will be a rational, non-anthropic and falsifiable explanation for the values of the parameters we observe.

- We shall have the basic framework of the quantum theory of gravity by 2010, 2015 at the outside. The last step will be the discovery of how to reformulate Newton's principle of inertia in the language of a quantum spacetime. It will take many more years to work out all the consequences, but the basic framework will be so compelling and natural as to remain fixed, once it is discovered.

- Within ten years of having the theory new kinds of experiment will be invented which will be able to test it. And the quantum theory of gravity will make predictions about the early universe which will be tested by observations of radiation from the big bang, including the cosmic microwave background radiation and gravitational radiation.

- By the end of the twenty-first century, the quantum theory of gravity will be taught to high-school students all around the world.

I started *Three Roads to Quantum Gravity* in the fall of 1999 and sent off final corrections to the publisher in October 2000. Since then the field has seen very dramatic progress on the road to quantum gravity.

The most exciting development is the possibility of observing the atomic structure of space itself. I mentioned this possibility briefly at the end of Chapter 10. Now there are even stronger indications that the atomic structure of space can be observed by current experiments. Indeed, Giovanni Amelino-Camelia and Tsvi Piran have pointed out that such observations may already have occurred.

These new observations are potentially as important as any that have occurred in the history of physics, for if they mean what some of us believe they mean, they mark the end of one era and the beginning of another.

Vast as it is, our universe is nowhere near empty. Where there is nothing else, there is radiation. We know of several different forms of radiation that travel in the spaces between the galaxies. One of them consists of very energetic particles, which we call *cosmic rays*. These appear to be mainly protons, with a mixture of heavier particles. Their distribution on the sky is uniform, which suggests they come from outside our galaxy. Scientists have observed these cosmic rays hitting the Earth's atmosphere with energies more than 10 million times the force achievable by the largest particle accelerators.

These cosmic rays are said to originate in highly energetic events in the centers of certain galaxies, which serve as a kind of natural particle accelerator. The rays come from regions of huge magnetic fields, perhaps produced by a supermassive black hole. Such things were once the stuff of fantasy, but we have more and more evidence for their existence. Although there are still uncertainties in our understanding of the origins of the cosmic rays, it seems most probable that the most energetic ones come from far outside our galaxy.

Consider then the most energetic cosmic ray protons observed, traveling toward us from a distant galaxy. At the energies they are traveling, about 10^{10} times the energy of the proton, or more than 10 million times the energy of the largest human-made particle accelerator, they are traveling very, very close to the speed of light. As our proton travels, it encounters another form of radiation that fills the space between galaxies—the *cosmic microwave background.*

The cosmic microwave background is a bath of microwaves that we understand as vestiges left over from the big bang. This radiation has been observed to come at us equally from all directions, up to small deviations of around a few parts in a hundred thousand. It now has a temperature of 2.7 degrees above absolute zero, but it was once at least as hot as the center of a star and cooled to its present temperature as the universe expanded. Given how uniformly we observe it to come from all directions of the universe, it is inconceivable that this radiation does not fill all space.

As a consequence, we know that our cosmic ray proton will encounter many photons from the microwave background as it travels through space. Most of the time nothing happens as a result of these interactions, because the cosmic ray proton has so much more energy and momentum than the photon it encounters. But if the proton has enough energy, it sometimes produces another elementary particle. When this happens, the cosmic ray slows down and loses energy because it takes energy to create the new particle.

The lightest particle that can be created in this way is called a *pion*. Using the basic laws of physics, including Einstein's special theory of relativity, one can work out a simple prediction about the processes by which cosmic ray protons and photons from the cosmic microwave background interact to make pions. The prediction is that there is a certain energy—called a *threshold*—above which this is very likely to happen. A proton above this energy will continue to interact in this way, losing energy each time, until it is slowed down enough that its energy falls below the threshold.

This works something like a 100% tax. Suppose there were some income, say $1 billion, above which all income would be taxed at 100%. Then no one would ever earn above $1 billion a year, because 100% of their income above this amount would be taxed. Our case is like a 100% tax on energy, as all the energy that a cosmic ray proton may have above the threshold will be removed, through processes that produce pions by its interacting with the cosmic microwave background.

This formula dictates that cosmic ray protons cannot hit the earth with an energy greater than the threshold energy. There is ample time in the protons' journey for any additional energy to be siphoned off in creating multiple pions.

I want to emphasize that this formula derives from the well-tested laws of special relativity—the results should therefore be very reliable. Thus, when this prediction was proposed by three Russian physicists with the names of Greisen, Zatsepin, and Kuzmin in the 1960s, it was very well received in the scientific community. Researchers had no reason to believe that cosmic ray protons would ever be seen with energies greater than the threshold.

Convincing as it was, Greisen, Zatsepin, and Kuzmin's prediction turned out to be wrong. In the last several years, many cosmic rays have been seen with energies greater than the threshold. This startling piece of news has galvanized scientists in the field. It is called the Ultra High Energy Cosmic Ray, or UHECR, anomaly.

Three explanations have been proposed for this effect. The first is astrophysical, and suggests that cosmic rays, or at least those above the threshold, are produced inside our galaxy, close enough that the effect may not have removed all their energy. The second solution is physical, and posits that the particles making up the very high energy cosmic rays are not protons, but actually much heavier particles, which do not lose energy by interacting with the microwave background. Instead, they decay over time, giving rise to the protons we observe. However, their lifetime is hypothesized to be extremely long, so that they are able to travel for many millions of years before they decay.

Both of these explanations appear far-fetched. There is no evidence for either nearby sources of cosmic rays or such heavy meta-stable particles. Moreover, both theories would require careful adjustments of parameters to unusual values just to fit these observations.

The third explanation has to do with quantum gravity. The atomic structure predicted by loop quantum gravity, which I described in Chapters 9 and 10, is expected to modify the laws that govern the interactions of elementary particles. This modification has the effect of changing the location of the threshold, and it is very natural that the result may be to raise the threshold enough to explain all the observations so far made.

This explanation leads to new predictions. First, the threshold may be seen at higher energy, in new experiments that will be able to detect cosmic rays at still higher energies. This is not the case with the other two explanations. Second, the effect must be universal, as the quantum geometry of spacetime must affect all particles that move. Hence the same effect must be seen in other particles.

There is in fact one case in which a similar effect may have been observed. Very energetic busts of photons arrive on Earth. These busts are called *gamma ray busts* and *blazars*, and they are believed to originate far outside our galaxy and travel for billions of years before arriving on Earth. Their origin is controversial, but it is possible they are the result of collisions among neutron stars or black holes. The most energetic of these are subject to a threshold for a similar reason, because they may interact with a background of diffuse starlight coming from all the stars in the universe. As in the case of the cosmic rays, photons have been seen with energies that exceed that threshold, coming from an object called *Markarian 501*.

Thus, all of a sudden, there is a real possibility that quantum gravity has become an experimental science. This is the most important thing that could have happened. It means that experimental relevance, rather than individual taste or peer pressure, must now become the determining factor for the correctness of an idea about quantum gravity.

Moreover, in the last several months, a startling implication of the theory of quantum gravity has emerged. *This is the possibility that the speed of light may depend on the energy carried by a photon.* This effect appears to come about as a result of the interaction of light with the atomic structure of space. These effects are tiny and so do not contradict the fact that so far all observations have concluded that the speed of light is constant. But for photons that travel very long distances across the universe, they add up to a significant effect, which can be observed with current technology.

The effect is very simple. If higher frequency light travels slightly faster than lower frequency light, then if we observe a very short burst of light coming from very far away, the higher energy photons should arrive slightly before those of lower energy. This could be observed in the gamma ray busts. The effect has not yet been seen, but if it is indeed there, it may be observed in experiments planned for the near future.

At first I was completely shocked by this idea. How could it be right? Relativity, based on the postulate of the constancy of the speed of light, is the foundation of all our understanding of space and time.

But as some wiser people explained to me, these new developments do not necessarily contradict Einstein. The basic principles enunciated by Einstein, such as the relativity of motion, may remain true. There still is a universal speed of light, which is the speed of the least energetic photons. What these developments imply is that Einstein's insights must be deepened to take into account the quantum structure of space and time, just as Einstein deepened Descartes's and Galileo's insights about the relativity of motion. It may be time for us to add another layer of insight into our understanding of what motion is.

Exactly how relativity is to be modified is a subject of hot debate at the moment. Some people argue that special relativity theory must be modified to account for the atomic structure of spacetime predicted by loop quantum gravity. According to loop quantum gravity, all observers see the discrete structure of space below the planck length. This seems to contradict relativity, which tells us that lengths are measured differently by different observers—the famous length contraction effect. One resolution is that special relativity can be modified so that there is one length scale, or one energy scale, that all observers agree on. Thus, while all other lengths will be measured differently by different observers, for the special case of the Planck length all observers will agree. There is still complete relativity of motion, as posited by Galileo and Einstein. But one consequence is that the speed of light can pick up a small dependence on energy.

I heard about the possibility of such a new twist on relativity from several people at once: Giovanni Amelino-Camelia, Jurek Kowalski-Glikman, and Joao Magueijo. At first I told them this was the craziest thing I'd ever heard, but Joao, who was my colleague in London at the time, was patient enough to keep coming back many times, until I finally got it. Since then I've seen other people go through this process. It is interesting to observe one of Thomas Kuhn's famous paradigm shifts in action.

Another hot topic is whether the possible variation of the speed of light with energy has consequences for our understanding of the history of the universe. Suppose that the speed of light increases with energy. (This is not the only possibility, but it is so far allowed by the observations we have.) When the universe was in its early stages, then, the average speed of light would have been higher, because the universe was then very hot, and hot photons have more energy. This idea has the possibility of solving a number of puzzles that cosmologists are very concerned about. For example, we don't know why the temperature in early times was nearly the same everywhere in the universe, in spite of the fact that there had not yet been time for all the regions to interact with one another. If the speed of light is higher than we currently think, there may have been time for all parts of the universe to have been in contact, and the mystery is solved! Indeed, cosmologists such as Andrew Albrecht and Joao Magueijo had already speculated about this possibility.

These puzzles have inspired a theory called *inflation*, which posits that the universe expanded at an exponentially increasing rate during a short period very early in its history. This theory has had some successes, but there have remained open questions about its connection with the more fundamental theory—the theory of quantum gravity. It is fascinating that a new idea has emerged based on our theories of quantum gravity, which may address this puzzle. This is good, because it is a spur to new observations that may decide which solution is right. It is often easier to use experiment to choose between two competing theories than it is to demonstrate that a single theory is right or wrong. Of course, the experiments may show instead that some combination of the two theories is right.

But most important, new observations that give evidence for or against the effects of quantum gravity on the propagation of light offer the chance to prove the validity of the theories described in this book. String theory and loop quantum gravity, for example, are likely to make different predictions for the results of these experiments. Loop quantum gravity appears to require modifications in special relativity. String theory, on the other hand, at least in its simplest versions, assumes that special relativity remains true no matter how small the distances are probed.

This is good news indeed, for as soon as the light of experiment is turned on, sociological forces such as govern academic politics and fashion must slink back to the shadows, as the judgement of nature supercedes the judgements of professors.

This is not the only place where cosmological observation and fundamental theory are confronting each other. An even more exciting—and for some disturbing—case has to do with the *cosmological constant*. This refers to the possibility—first realized by Einstein—that empty space might have a non-zero energy density. This energy density would be observable in the effect it has on the expansion of the universe.

Once this possibility was accepted, it led to a major crisis in theoretical physics. The reason is that the most natural possibility allowed for the value of this empty space energy density is that it should be huge—more than a hundred powers of ten larger than is compatible with observation. The exact value—which is what the name *cosmological constant* refers to—cannot be predicted by current theory. In fact, we can adjust a parameter to get any value for the cosmological constant we want. The problem is that to avoid a huge cosmological constant, the parameter has to be adjusted to an accuracy of at least 120 decimal places. How such a precise adjustment is to be obtained is a mystery.

This is perhaps the most serious problem facing fundamental physics, and it recently got worse. Until a few years ago, it was almost universally believed that even if it required a very precise adjustment, in the end the cosmological constant would be exactly zero. We had no idea why the cosmological constant would be zero, but at least zero is a simple answer. However, recent observations have suggested that the cosmological constant is not zero; it has instead a very small, but positive, value. This value is tiny on the scales of fundamental physics; in Planck units it is around 10^{-120} (or .0000. . . .) with 120 zeros before one encounters a non-zero digit.

But even though tiny when measured in fundamental units, this value is large enough to have a profound effect on the evolution of our universe. This cosmological constant would make the energy density of empty space equal to about twice the current value of the energy density of everything else that has been observed. This may seem surprising, but the point is that the energy density of all the kinds of matter that have been observed is currently very small. This is because the universe is very old. When measured in fundamental units, its present age is about 10^{60} Planck times. And it has been expanding all this time, thus diluting the density of matter.

The energy density due to the cosmological constant does not, as far as we know, dilute as the universe expands. This gives rise to a very troubling question: Why is it that we live at a time when the matter density has diluted to the point that it is of the same order of magnitude as the density due to the cosmological constant?

I do not know the answer to any of these questions. Neither, I think, does anyone else, although there are a few interesting ideas on the table.

However, the apparent fact that the cosmological constant is not zero has big implications for the quantum theory of gravity. One reason is that it seems to be incompatible with string theory. It turns out that a mathematical structure that is required for string theory to be consistent—which goes by the name *supersymmetry*—only permits the cosmological constant to exist if it has the opposite sign from the one that has apparently been observed. There are some interesting studies of string theory in the presence of a negative cosmological constant, but no one so far knows how to write down a consistent string theory when the cosmological constant is positive—as has apparently been observed.

I do not know if this obstacle will kill string theory—string theorists are very resourceful, and they have often expanded the definition of string theory to include cases once thought impossible. But string theorists are worried, for if string theory cannot be made compatible with a positive cosmological constant—and that continues to be what the astronomers observe—then the theory is dead.

But there is a second reason why a positive cosmological constant is troubling for quantum theories of gravity, including string theory. As the universe continues to expand, the energy density due to matter will continue to dilute. But the cosmological constant is believed to remain stable. This means that there will be a time in the future when the cosmological constant comprises most of the energy density in the universe. After this the expansion will accelerate—indeed the effect is very similar to the inflation proposed for the very early universe.

To be an observer in an inflating universe is to be in a very poor situation. As the universe inflates, we will see less and less of it. Light cannot keep up with the acceleration of the expansion, and light from distant galaxies will no longer be able to reach us. It would be as if large regions of the universe had fallen behind the horizon of a black hole. One by one distant galaxies will go over a horizon, to a zone from which their light will never again reach us. With the value apparently measured, it is only a matter of a few tens of billions of years before observers in a galaxy see nothing around them except their own galaxy surrounded by a void.

In such a universe, the considerations of Chapters 1–3 become crucial. A single observer can only see a small portion of the universe, and that small portion will only decrease over time. No matter how long we wait, we will never see more of the universe than we do now.

Tom Banks has expressed this principle beautifully. There is a finite limit to the amount of information that any observer in an inflating universe may ever see. The limit is that each observer can see no more than $\frac{3\pi}{G^2L}$ bits of information, where G is Newton's constant and L is the cosmological constant. Raphael Bousso called this the *N-bound* and argued that this principle may be derived by an argument that is closely related to Bekenstein's bound, which is described in Chapters 8 and 12. The principle seems to be required by the second law of thermodynamics.

As the universe expands, we would expect that it contains more and more information. But, according to this principle, any given observer can only see a fixed amount of information given by the N-bound.

In this circumstance, the traditional formulations of quantum theory fail because they assume that an observer can, given enough time, see anything that happens in the universe. It seems to me that there is then no alternative but to adopt the program I described in Chapter 3, which was proposed by Fotini Markopoulou—to reformulate physics in terms of only what observers inside the universe can actually see. As a result, Markopoulou's proposal has been getting more attention from people on both sides of the string theory/loop quantum gravity divide.

So far there is no proposal for how to reformulate string theory in such terms. One possible step toward such a formulation is Andrew Strominger's new proposal, which applies the holographic principle to spacetimes, with a positive cosmological constant.

At the same time, loop quantum gravity is clearly compatible with such a reformulation of quantum theory—it is already background-independent and expressed in a language in which the causal structure exists all the way down to the Planck scale.

In fact, Bank's N-bound is easy to derive in loop quantum gravity, using the same methods that led to the description of the quantum states on black hole horizons. Moreover, in loop quantum gravity there is a complete description of a quantum universe filled with nothing but a positive cosmological constant. This is given by a certain mathematical expression, discovered by the Japanese physicist Hideo Kodama. Using Kodama's result, we are able to answer previously unsolvable questions, such as exactly how the solutions of Einstein's general relativity theory emerge from the quantum theory. Thus, at least in our present stage of knowledge, while string theory has trouble incorporating the apparently observed positive value of the cosmological constant, loop quantum gravity seems to prefer that case.

Beyond this, there has continued to be steady progress in loop quantum gravity. The work of two young physicists,

Chopin Soo and Martin Bojowald, has led to a greatly improved understanding of how classical cosmology emerges from loop quantum gravity. New calculational methods for spin foams have given us very satisfactory results. Large classes of calculations, for example, turn out to give finite, well-defined answers, where conventional quantum theories gave infinities. These results present more evidence that loop quantum gravity provides a consistent framework for a quantum theory of gravity.

Before closing I want to emphasize again that this book describes science *in the making*. There are some people who think that popular science should be restricted to reporting discoveries that have been completely confirmed experimentally, leaving no room for controversy among experts. But restricting popular science in this way blurs the line between science and dogma, and dictates how we believe the public should think. To communicate how science really works, we must open the door and let the public watch as we go about searching for the truth. Our task is to present all the evidence and invite the readers to think for themselves.

But this is the paradox of science: It is an organized, even ritualized, community designed to support the process of a large number of people thinking for themselves and discussing and arguing the conclusions they come to.

Exposing the debates in a field like quantum gravity to the public is also bound to raise controversy among experts. In this book, I tried to treat the different approaches to quantum gravity as evenhandedly as possible. Still, some experts have told me I do not praise string theory enough, whereas others have told me I did not emphasize its shortcomings nearly enough. Some colleagues complained that I did not champion my own field of loop quantum gravity strongly enough, given that string theorists generally fail to even mention loop quantum gravity—or anything other than string theory—in their own books and public talks. Indeed, one string theorist who reviewed the book called me a "maverick" for even mentioning that many of the leading people who made key discoveries in quantum gravity did not work on string theory. I take the fact that this kind of criticism came from both sides as evidence that I did not completely fail to present an evenhanded view of the successes and failures of loop quantum gravity, string theory, and the other approaches to quantum gravity.

At the same time, I cannot help but notice that as time goes on, it appears that the close-mindedness that characterizes the thinking of some (of course, not all) string theorists does appear to have inhibited progress. Many string theorists seem disinterested in thinking about questions that cannot be sensibly posed within the existing framework for string theory. This is perhaps because they are convinced that supersymmetry is more fundamental than the lesson from general relativity that spacetime is a dynamical, relational entity. Nevertheless, I suspect this is the main reason for the slow progress on key questions such as making string theory background independent, or understanding the role of the dynamics of causal structure, problems that cannot be addressed without going beyond current string theory. Of course, other people can and do work on this problem, and we are making progress on it, even if we are not considered by the orthodox to be "real string theorists."

My own view remains optimistic. I believe that we have on the table all the ingredients we need to make the quantum theory of gravity and that it is mostly a matter of putting the pieces together. So far, nothing has changed my understanding that loop quantum gravity is a consistent framework for a complete quantum theory of spacetime, and string theory does not yet provide more than a background-dependent approximation to such a theory. I believe that some aspects of string theory might nevertheless play a role, as an approximation to the real theory, but given a choice between the two, loop quantum gravity is certainly the deeper and more comprehensive theory. Furthermore, if the atomic structure of spacetime predicted by loop quantum gravity requires modifications of special relativity such as a variation in the speed of light with energy, this is a challenge for string theory, which in its current form assumes the theory makes sense without such effects. So if—as conjectured in Chapter 14—a form of string theory can be derived from loop quantum gravity, it may be in a modified form.

But what is important above all is that it doesn't matter what I or any other theorist thinks. Experiment will decide. And quite possibly in the next few years.

Lee Smolin
March 3, 2002
Waterloo, Canada

....

Terms in italics have their own glossary entries.

absolute space and time
Newton's view of space and time according to which they exist eternally, independent of whether anything is in the universe or not and of what happens inside the universe.

angular momentum
A measure of rotational motion, analogous to momentum. The total angular momentum of an isolated system is conserved.

background
A scientific model or theory often describes only part of the universe. Some features of the rest of the universe may be included as necessary to define the properties of that part of the universe that is studied. These features are called the background. For example, in Newtonian physics space and time are part of the background because they are taken to be absolute.

background dependent
A theory, such as Newtonian physics, that makes use of a *background*.

background independent
A theory that does not make use of a division of the universe into a part that is modelled and the rest, which is taken to be part of the *background*. General relativity is said to be background independent because the geometry of space and time is not fixed, but evolves in time just as any other field, such as the electromagnetic field.

Bekenstein bound
The relationship between the area of a surface and the maximum amount of information about the universe on one side of it that can pass through it to an observer on the other side. The relationship states that the number of bits of information the observer can gain

cannot be greater than one-quarter the area of the surface in Planck units.

black hole

A region of space and time that cannot send signals to the outside world because all light emitted comes back. Among the ways a black hole may be formed is by the collapse of a very massive star when it runs out of its nuclear fuel.

black hole horizon

The surface surrounding a *black hole*, within which is the region from which light signals cannot escape.

boson

A particle whose angular momentum comes in integer multiples of *Planck's constant*. Bosons do not obey the *Pauli exclusion principle*.

brane

A possible feature of geometry, as described in *string theory*, which consists of a surface of some dimensions embedded in space, which evolves in time. For example, strings are one-dimensional branes.

causality

The principle that events are influenced by those in their past. In relativity theory one event can have a causal influence on another only if energy or information sent from the first reaches the second.

causal structure

Because there is a maximum speed at which energy and information can be transmitted, the events in the history of the universe can be organized in terms of their possible causal relations. To do this one indicates, for every pair of events, whether the first is in the causal future of the second, or vice versa, or whether there is no possible causal relation between them because no signal could have travelled between them. Such a complete description defines the causal structure of the universe.

classical theory

Any physical theory that shares certain features with *Newtonian physics*, including the assumption that the future is completely determined by the present and that the act of observation has no effect on the system studied. The term is used mainly to label any theory that is not part of quantum theory. Einstein's general theory of relativity is considered to be a classical theory.

classical physics

The collection of *classical theories*.

consistent histories

An approach to the interpretation of quantum theory which asserts that the theory makes predictions about the probabilities for sets of alternative histories, when these can be done consistently.

continuous

Describing a smooth and unbroken space which has the property of the number line, which is that it can be quantified in terms of coordinates expressed in *real numbers*. Any region of continuous space having a finite volume contains an infinitely uncountable number of points.

continuum

Any space that is *continuous*.

curvature tensor

The basic mathematical object in Einstein's *general theory of relativity*. It determines how the tipping of *light cones* changes from time to time and place to place in the history of the universe.

degree of freedom

Any variable in a physical theory that may be specified independently of the other variables, which once specifies evolves in time according to a dynamical law. Examples are the positions of particles and the values of the electric and magnetic fields.

diffeomorphism

An operation that moves the points of space around, preserving only those relationships between them that are used to define which points are near to one another.

discrete

Describing a space that is made of a finite number of points.

duality

The principle of duality applies when two descriptions are different ways of looking at the same thing. In particle physics it usually refers to a description in terms of strings and a description in terms of the flux of the electric field or some generalization of it.

Einstein equations

The basic equations of the *general theory of relativity*. They determine how *light cones* tip and how they are related to the distribution of matter in the universe.

electromagnetism

The theory of electricity and magnetism, including light, developed by Michael Faraday and James Clerk Maxwell in the nineteenth century.

entropy

A measure of the disorder of a physical system. It is defined as the amount of *information* about the microscopic motion of the atoms making up the system which is not determined by a description of the macroscopic state of that system.

equilibrium

A system is defined to be in equilibrium, or thermodynamic equilibrium, when it has the maximum possible amount of *entropy*.

event

In relativity theory, something that happens at a particular point of space and moment of time.

exclusion principle

see *Pauli exclusion principle*.

fermion

A particle whose angular momentum comes in integer multiples of one-half of *Planck's constant*. Fermions satisfy the *Pauli exclusion principle*.

Feynman diagram

A depiction of a possible process in the interaction of several elementary particles. Quantum theory assigns to each diagram the probability amplitude for that process to occur. The total probability is proportional to the square of the sum of the amplitudes of the possible processes, each of which is depicted by a Feynman diagram.

field

A physical entity that is described by specifying the value of some quantity at every point of space and time; examples are the electric and magnetic fields.

future

The future, or causal future, of an event consists of all those events that it can influence by sending energy or information to it.

future light cone

For a specific event, all other events that can be reached from it by a signal travelling at the speed of light. Since the speed of light is the maximum speed at which energy or information can travel, the future light cone of an event marks the limits of the causal *future* of that event. See also *light cone*.

general theory of relativity

Einstein's theory of gravity, according to which gravity is related to the influence the distribution of matter has on the *causal structure* of spacetime.

graph

A diagram consisting of a set of points, called vertices, connected by lines, called edges. See also *lattice*.

Hawking radiation

The thermal radiation *black holes* are predicted to give off, having a temperature which is inversely proportional to the black hole's mass. Hawking radiation is caused by quantum effects.

hidden variables

Conjectured degrees of freedom which underlie the statistical uncertainties in quantum theory. If there are hidden variables, then it is possible that the uncertainties in quantum theory are just the result of our ignorance about the values of the hidden variables and are not fundamental.

horizon

For each observer in a *spacetime*, the surface beyond which they cannot see, or receive any signals from. Examples are *black hole horizons*.

information

A measure of the organization of a signal. It is equal to the number of yes/no questions whose answers could be coded in the signal.

knot theory

A branch of mathematics concerned with classifying the different ways of tying a knot.

lattice

A space consisting of a finite number of points, with nearby points connected by lines called edges. A lattice is often, but not always, distinguished from a *graph* in that a lattice is a graph with a regular structure. An example of a lattice is shown in Figure 22.

lattice theory

A theory in which space or *spacetime* is considered to be a *lattice*.

light cone

All the events that can be reached by light signals travelling to the future, or coming from the past, from a single event. We may therefore distinguish between the *future light cone*, which contains events that can be reached by light travelling into the future, and the *past light cone*, which contains events that can be reached by light travelling from the past.

link

Two curves link in three-dimensional space if they cannot be pulled apart without passing one through the other.

loop

A circle drawn in space.

loop quantum gravity

An approach to quantum gravity in which space is constructed from the relationships between loops, originally derived by applying quantum theory to the formulation of general relativity discovered by Sen and Ashtekar.

many-worlds interpretation

An interpretation of quantum theory according to which the different possible outcomes of an observation of a quantum system reside in different universes, all of which somehow coexist.

M theory

The conjectured theory which would unify the different string theories.

Newton's gravitational constant

The fundamental constant that measures the strength of the gravitational force.

Newtonian physics

All physical theories formulated on the pattern of Newton's laws of motion. See *classical physics*, which is a synonymous term.

non-commutative geometry

A description of a space in which it is impossible to determine enough information to locate a point, but which can have many other properties of space including the fact that it can support a description of particles and fields evolving in time.

past *or* **causal past**

For a particular event, all other events that could have influenced it by sending energy or information to it.

past light cone

The past light cone of an event consists of all those events that could have sent a light signal to it.

Pauli exclusion principle

The principle that no two fermions can be put into exactly the same *quantum state*; named after Wolfgang Pauli.

perturbation theory

An approach to making calculations in physics in which some phenomena are represented in terms of small deviations from or oscillations of some stable state, or the interactions among such oscillations.

Planck scale

The scale of distance, time and energy on which quantum gravity effects are important. It is defined roughly by the *Planck units* – processes on the Planck scale take around a *Planck time*, which is 10^{-43} of a second. To observe on the Planck scale, distances of around the *Planck length* must be probed. This is about 10^{-33} of a centimetre.

Planck's constant

A fundamental constant that sets the scale of quantum effects; normally denoted by h.

Planck units

The basic units of measure in a quantum theory of gravity. Each is given by a unique combination of three basic constants: *Planck's constant, Newton's gravitational constant* and the speed of light. Planck units include the Planck length, Planck energy, Planck mass, Planck time and Planck temperature.

quantum chromodynamics (QCD)

The theory of the forces between *quarks*.

quantum electrodynamics (QED)

The marriage of *quantum theory* with electrodynamics. It describes light and the electric and magnetic forces in quantum terms.

quantum cosmology

The theory that attempts to describe the whole universe in the language of *quantum theory.*

quantum gravity

The theory that unifies quantum theory with Einstein's general theory of relativity.

quantum theory *or* **quantum mechanics**

The theory of physics that attempts to explain the observed behaviour of matter and radiation. It is based on the *uncertainty principle* and *wave–particle duality.*

quantum state

The complete description of a system at one moment of time, according to the *quantum theory.*

quark

An elementary particle which is a constituent of a proton or neutron.

real number

A point on the *continuous* number line.

relational

Describing a property that describes a relationship between two objects.

relational quantum theory

An interpretation of quantum theory according to which the *quantum state* of a particle, or of any subsystem of the universe, is defined, not absolutely, but only in a context created by the presence of an observer, and a division of the universe into a part containing the observer and a part containing that part of the universe from which the observer can receive information. Relational quantum cosmology is an approach to quantum cosmology which asserts that there is not one quantum state of the universe, but as many states as there are such contexts.

relativity theory

Einstein's theory of space and time, comprising the special theory of relativity, which describes the *causal structure* of *spacetime* without gravity, and the *general theory of relativity*, in which the causal structure becomes a dynamical entity that is partly determined by the distribution of matter and energy.

second law of thermodynamics

The law stating that the *entropy* of an isolated system can only increase in time.

spacetime

The history of a universe, comprising all its events and their relationships.

speed of light

The speed at which light travels, which is known to be the maximum speed for the transmission of energy and of information.

spin

The angular momentum of an elementary particle which is an intrinsic property of it, independent of its motion.

spin network

A graph whose edges are labelled by numbers representing spins. In *loop quantum gravity* each quantum state of the geometry of space is represented by a spin network.

spontaneous symmetry breaking

The phenomena by which a stable *state* of a system can have less *symmetry* than the laws that govern the system.

state

In any physical theory, the configuration of a system at a specified moment of time.

string

In *string theory*, the basic physical entity, the different states of which represent the different possible elementary particles. A string can be visualized as a path or a loop that propagates through a *background* space.

string theory

A theory of the propagation and interactions of *strings*, in *background* spacetimes.

supersymmetry

A conjectured *symmetry* of elementary particle physics and *string theories* which asserts that *bosons* and *fermions* exist in pairs, each member of which has the same mass and interactions.

supergravity

An extension of Einstein's *general theory of relativity* in which the different kinds of elementary particle are related to one another by one or more *supersymmetries*.

symmetry

An operation by which a physical system may be transformed without affecting the fact that it is a possible state or history of the system. Two *states* connected by a symmetry have the same energy.

temperature

The average kinetic energy of a particle or mode of vibration in a large system.

thermal *or* thermodynamic equilibrium

See *equilibrium*.

topos theory

A mathematical language which is appropriate for describing theories in which properties are context dependent, as in *relational quantum theory*.

twistor theory

An approach to quantum gravity invented by Roger Penrose in which the primary elements are causal processes and the events of spacetime are constructed in terms of the relationships between the causal processes.

uncertainty principle

A principle in *quantum theory* according to which it is impossible to measure both the position and momentum (or velocity) of a particle or, more generally, the state and rate of change of any system.

wave–particle duality

A principle of *quantum theory* according to which one can describe elementary particles as both particles and waves, depending on the context.

SUGGESTIONS FOR FURTHER READING

Here I give a brief list of sources where the interested reader can find more information about the topics discussed. More information will be available on a Website, http://www.qgravity.org.

INTRODUCTION AND POPULAR TEXTS

Many books aim to introduce the reader to the basic ideas of quantum theory and general relativity. They cater to all different levels, from comic books and children's books to philosophical treatises. There are so many that the reader is advised to go to the science section of a good bookshop, look at the various books on quantum theory and relativity, read the first few pages of each and take the one you like best. The reader may also find it interesting to look at the popularizations by the inventors of these theories: Bohr, Einstein, Heisenberg and Schrödinger have all written introductions to their work for the layperson.

My own *Life of the Cosmos* (Oxford University Press, New York and Weidenfeld & Nicolson, London, 1996) introduces the basic ideas of quantum theory and general relativity in Parts 4 and 5.

Brian Greene's *The Elegant Universe* (Norton, 1999) gives a very good introduction to the basic ideas of string theory and the problems it currently faces. Roger Penrose's books, especially the *Emperor's New Mind* (Oxford University Press, 1989), are a good introduction to the problem of quantum gravity and quantum black holes, emphasizing of course his own point of view.

REFERENCE TO THE SCIENTIFIC LITERATURE
....

Virtually the whole of the scientific literature on topics relevant to theoretical physics since 1991 is available in an electronic archive, which can be found at http://xxx.lanl.gov/. Note that while you generally have to have a professional affiliation to publish at this site, anyone can download and read the articles archived there. The papers of relevance to this book are mostly found in the archives hep-th and gr-qc. A search for the people mentioned below will return a list of the papers which underlie the developments described.

Another very good source for the ideas and mathematical developments used in quantum gravity is John Baez's Website, This Week's Finds in Mathematical Physics, at http://math.ucr.edu/home/baez/ TWF.html. He also has a nice online tutorial introduction to general relativity at http://math.ucr.edu/home/baez/gr/gr.html. The reader wanting a general introduction to the history of quantum gravity and its basic issues may find the following articles interesting: Carlo Rovelli, 'Notes for a brief history of quantum gravity', gr-qc/0006061; Carlo Rovelli, 'Quantum spacetime – what do we know?', gr-qc/ 9903045, and Lee Smolin, 'The new universe around the next corner', in *Physics World*, December 1999.

Most of the following key references are in the xxx.lanl.gov archive. A more complete list of references is available at the Website mentioned above.

CHAPTER 2

The discussion of the logic of observers inside the universe is based on F. Markopoulou, 'The internal description of a causal set: What the universe looks like from the inside', gr-qc/9811053, *Commun. Math. Phys.* **211** (2000) 559–583.

CHAPTER 3

The consistent histories interpretation is described in R.B. Griffiths, *Journal of Statistical Physics* **36** (1984) 219; R. Omnes, *Journal of Statistical Physics* **53** (1988) 893; and M. Gell-Mann and J.B. Hartle in *Complexity, Entropy, and the Physics of Information*, SFI Studies in the Sciences of Complexity, Vol. VIII, edited by W. Zurek (Addison Wesley, Reading, MA, 1990). The criticisms of Kent and Dowker are found in Fay Dowker and Adrian Kent, 'On the consistent histories approach to quantum mechanics', *Journal of Statistical Physics.* **82** (1996) 1575. Gell-Mann and Hartle comment in 'Equivalent sets of histories and multiple quasiclassical realms', gr-qc/9404013; J. B. Hartle, gr-qc/

9808070. The reformulation of the consistent histories formulation in terms of topos theory, which emphasizes its relational aspects, is found in C.J. Isham and J. Butterfield, 'Some possible roles for topos theory in quantum theory and quantum gravity', gr-qc/9910005. Other relational approaches to quantum cosmology are found in L. Crane, *Journal of Mathematical Physics* **36** (1995) 6180; L. Crane, in *Knots and Quantum Gravity*, edited by J. Baez (Oxford University Press, New York, 1994); L. Crane, 'Categorical physics', hep-th/9301061; F. Markopoulou, 'Quantum causal histories', hep-th/9904009, *Class. Quan. Grav.* **17** (2000) 2059–2072; F. Markopoulou, 'An insider's guide to quantum causal histories', hep-th/9912137, *Nucl. Phys. Proc. Suppl.* **88** (2000) 308–313; C. Rovelli, 'Relational quantum mechanics', quant-ph/9609002, *International Journal of Theoretical Physics* **35** (1996) 1637; L. Smolin, 'The Bekenstein bound, topological field theory and pluralistic quantum cosmology', gr-qc/950806.

CHAPTER 4
The process formulation of quantum theory was developed first by David Finkelstein, whose work is the main inspiration for this chapter. It is described in David Ritz Finkelstein, *Quantum Relativity: A Synthesis of the ideas of Einstein and Heisenberg* (Springer-Verlag, 1996). Rafael Sorkin has also pioneered the exploration of the role of causality in quantum gravity.

CHAPTERS 5–8
This is all standard material in classical general relativity and quantum field theory. Good introductions are N.D. Birrell and P.C.W. Davies, *Quantum Fields in Curved Spacetime* (Cambridge University Press, 1982); and Robert M. Wald, *Quantum Field Theory in Curved Spacetime and Black Hole Thermodynamics* (University of Chicago Press, 1994).

CHAPTERS 9 AND 10
There are several expositions of loop quantum gravity at a semi-popular or semi-technical level. They include Carlo Rovelli, 'Loop quantum gravity', gr-qc/9710008, Carlo Rovelli, 'Quantum spacetime: what do we know?', gr-qc/9903045; L. Smolin in *Quantum Gravity and Cosmology*, edited by Juan Perez-Mercader *et al.* (World Scientific, 1992); L. Smolin, 'The future of spin networks', in *The Geometric Universe* (1997), edited by S.A. Huggett *et al.* (Oxford University Press, 1998), gr-qc/9702030. The book by Rodolfo Gambini and Jorge Pullin, *Loops, Knots, Gauge Theories and Quantum Gravity* (Cambridge University Press, 1996) describes their approach to the subject.

The mathematically rigorous approach to loop quantum gravity is presented in Abhay Ashtekar, Jerzy Lewandowski, Donald Marolf, Jose Mourao and Thomas Thiemann, 'Quantization of diffeomorphism invariant theories of connections with local degrees of freedom', *Journal of Mathematical Physics* **36** (1995) 6456, gr-qc/9504018; Abhay Ashtekar, Jerzy Lewandowski, 'Quantum field theory of geometry', hep-th/9603083; and T. Thiemann, 'Quantum spin dynamics I and II', gr-qc/9606089, gr-qc/9606090, *Classical and Quantum Gravity* **15** (1998) 839, 875.

The original references for the Ashtekar–Sen formalism are in A. Sen, *Physics. Letters* **B119** (1982) 89; *International Journal; of Theoretical Physics* **21** (1982) 1; A. Ashtekar, *Physical Review Letters* **57** (1986) 2244; A. Ashtekar, *Physical. Review* **D36** (1987) 1587.

CHAPTER 11
This is all standard material in string theory, to which Brian Greene's *The Elegant Universe* (Norton, 1999) is an excellent introduction. The best textbook is J. Polchinksi, *String Theory* (Cambridge University Press, 1998).

CHAPTER 12
The original references for the holographic principle are Gerard 't Hooft, 'Dimensional reduction in quantum gravity', gr-qc/9310006, in *Salanfestschrift*, edited by A. Alo, J. Ellis, S. Randjbar-Daemi (World Scientific, 1993); and Leonard Susskind, 'The world as a hologram', hep-th/9409089, Journal of Mathematical Physics **36** (1995) 6377. Ideas closely related to the holographic principle were presented earlier by L. Crane in 'Categorical physics', hep-th/9301061 and hep-th/9308126 in *Knots and Quantum Gravity*, edited by J. Baez (Oxford University Press, 1994); L. Crane, 'Clocks and categories: is quantum gravity algebraic?' *Journal of Mathematical Physics* **36** (1995) 6180, gr-qc/9504038.

The Bekenstein bound was proposed in J.D. Bekenstein, *Lettere Nuovo Cimento* **4** (1972) 737, *Physical Review* **D7** (1973), 2333; *Physical Review* **D9** (1974) 3292. Ted Jacobson's paper deriving general relativity from the Bekenstein bound and the laws of thermodynamics is 'Thermodynamics of spacetime: the Einstein equation of state', gr-qc/9504004, *Physical Review Letters* **75** (1995) 1260. The derivation of the Bekenstein bound in loop quantum gravity is in L. Smolin, 'Linking topological quantum field theory and nonperturbative quantum gravity', gr-qc/9505028, *Journal of Mathematical Physics* **36** (1995) 6417. Another very promising version of the holographic principle was proposed by Rafael Bousso in 'A covariant entropy conjecture', hep-th/9905177, *Journal of High-Energy Physics,* **9907** (1999) 0004; R. Bousso, 'Holography in general space-times', hep-th/9906022, *Journal of High-Energy Physics* **9906** (1999) 028. A related theorem was proved in E. Flanagan, D. Marolf and R. Wald, hep-th/9908070. F. Markopoulou and I proposed a background independent version in 'Holography in a quantum spacetime', hep-th/9910146. In 'The strong and weak holographic principles', hep-th/0003056 I review the arguments for and against the different versions of the principle.

CHAPTER 13

The view of the relationship between loop quantum gravity and string theory is based on L. Smolin, 'Strings as perturbations of evolving spin networks', hep-th/9801022; L. Smolin, 'A candidate for a background independent formulation of *M* theory', hep-th/9903166; L. Smolin, 'The cubic matrix model and a duality between strings and loops', hep-th/006137.

There is an extensive literature on black holes in both string theory and loop quantum gravity. A sample of string theory papers is: A. Strominger and C. Vafa, *Physics Letters* **B379** (1996) 99, hep-th/9601029; C.V. Johnson, R.R. Khuri and R.C. Myers, *Physics Letters* **B378** (1996) 78, hep-th/9603061; J.M. Maldacena and A. Strominger, *Physical Review Letters* **77** (1996) 428, hep-th/9603060; C.G. Callan and J.M. Maldacena, *Nuclear Physics* **B472** (1996) 591, hep-th/9602043; G.T. Horowitz and A. Strominger, *Physical Review Letters* **77** (1996) 2368, hep-th/9602051.

A sample of papers on black holes in loop quantum gravity is: Carlo Rovelli, 'Black hole entropy from loop quantum gravity', gr-qc/9603063, *Physical Review Letters* **77** (1996) 3288; Marcelo Barreira, Mauro Carfora and Carlo Rovelli, 'Physics with nonperturbative quantum gravity: radiation from a quantum black hole', gr-qc/9603064, *General Relativity and Gravity* **28** (1996) 1293; Kirill Krasnov, 'On quantum statistical mechanics of a Schwarzschild black hole', gr-qc/9605047, *General Relativity and Gravity* **30** (1998) 53; Kirill Krasnov, 'Quantum geometry and thermal radiation from black holes', gr-qc/9710006, *Classical and Quantum Gravity* **16** (1999) 563; A. Ashtekar, J. Baez and K. Krasnov, 'Quantum geometry of isolated horizons and black hole entropy', gr-qc/0005126; A. Ashtekar, J. Baez, A. Corichi and K. Krasnov, 'Quantum geometry and black hole entropy', gr-qc/9710007, *Physical Review Letters* **80** (1998) 904.

Non-commutative geometry is introduced in the book by Alain Connes, *Non-commutative Geometry* (Academic Press, 1994).

CHAPTER 14
The material described here is mostly related to my book, *Life of the Cosmos*. The discussion of space is drawn from S. Kauffman and L. Smolin, 'Combinatorial dynamics in quantum gravity', hep-th/9809161.

Born in New York City, Lee Smolin was educated at Hampshire College and Harvard University. He is currently professor of Physics at the Center for Gravitational Physics for a unification of quantum theory, cosmology and relativity. He is the author of *The Life of the Cosmos* (1997).

$Ent. = \dfrac{Heat}{\infty}$

Dr. Steven E Nissen AMA/
   apolipoprotein
To remove bad colestrol
      11/5/03